Lecture Notes in Electrical Engineering

Volume 84

For further volumes:
http://www.springer.com/series/7818

Manuel Duarte Ortigueira

Fractional Calculus for Scientists and Engineers

Springer

Manuel Duarte Ortigueira
Faculdade de Ciências/Tecnologia da UNL
UNINOVA and DEE
Campus da FCT Quinta da Torre
2829-516 Caparica
Portugal
e-mail: mdo@fct.unl.pt

ISSN 1876-1100 e-ISSN 1876-1119

ISBN 978-94-007-0746-7 e-ISBN 978-94-007-0747-4

DOI 10.1007/978-94-007-0747-4

Springer Dordrecht Heidelberg London New York

Cover design: eStudio Calamar, Berlin/Figueres

Printed on acid-free paper

Springer is part of Springer Science+Business Media (www.springer.com)

To the memory of my parents

José Ortigueira and Rita Rosa da Ascenção Duarte

that did their best for me.

To my children

Eduardo and Joana, and my wife, Laurinda, for their love and patience.

Preface

Fractional Calculus has been attracting the attention of scientists and engineers from long time ago, resulting in the development of many applications. Since the nineties of last century fractional calculus is being rediscovered and applied in an increasing number of fields, namely in several areas of Physics, Control Engineering, and Signal Processing. However, most of the theoretical setup of Fractional Calculus was done by mathematicians that directed their attention preferably to the so-called Riemann-Liouville and/or Caputo derivatives. We must remark that most of the articles that appear in the scientific literature, in the framework of the fractional calculus and their applications, the authors use those derivatives but at the end they contrast their model using a numerical approach based in a finite number of terms from the series that define the Grünwald-Letnikov derivative. This may be confirmed in several books that appeared recently and is one justification for the present one. It intends to present a Fractional Calculus foundation based of the Grünwald-Letnikov derivative, because it exhibits great coherence allowing us to deduce from it the other derivatives, which appear as a consequence of the Grünwald-Letnikov derivative properties and not as a prescription. The Grünwald-Letnikov derivative is a straight generalisation of the classic derivative and leads to formulae and equations that recover the classic ones when the order becomes integer.

On the other hand, the available literature deals mainly with the causal (anti-causal) derivatives. In situations where no preferred direction exists it is common to use the Riesz potentials. Alternatively we will present the two-sided fractional derivatives that are more general than the Riesz potentials and are generalizations of the classic symmetric derivatives. These allow us to deal comfortably with fractional partial differential equations. Similarly, the Quantum Derivative is presented as a useful tool for dealing with the fractional Euler-Cauchy equations that are suitable for dealing with scale invariant systems. These derivatives are not described in published books.

This book is directed towards Scientists and Engineers mainly interested in applications who do not want to spend too much time and effort to access to the main Fractional Calculus features and tools. For this reason readers can "jump" the chapter 2 in a first reading. The book is written in a cursive way, like a divulgation text, reducing the formalism to increase the legibility. I hope I have been successful.

Acknowledgments

My travel through the FC world was made easier by the help I received from several people and I would like to refer here with a special thanks:

- Prof. Ferreira dos Santos from Instituto Superior Técnico, Lisbon, who taught me the beauty of complex variables functions and the two-sided Laplace transform.
- Prof. Steiger Garção President of the Electrical Engineering Department of the Faculty of Science and Technology and of UNINOVA by his incitements.
- Prof. José Sá da Costa from Instituto Superior Técnico, Lisbon, by his help in settling my things.
- My colleagues Fernando Coito, Arnaldo Batista, António Serralheiro, and Moisés Piedade by their friendship and discussions.

I would like to remember here and thank all the "integer" people I have been meeting at the Workshops and Symposia. Besides those I cite at the introduction I would like to refer, without any special order: Om Agrawal, Raoul Nigmatullin, Yury Luchko, Bohdan Datsko, Mohammad Tavazoei, Vladimir Uchaikin, Yang-Quan Chen, Jocelyn Sabatier, Pierre Melchior, Duarte Valério, Isabel de Jesus, and Francesco Mainardi. Probably I forgot others. I beg their pardon.

Contents

Chapter 1
A Travel Through the World of Fractional Calculus

I heard the designation "fractional derivative" for the first time in the 1980s. My curiosity was excited and I decided that later I would learn something about the subject. This only happened in 1994 when I decided to talk with Prof. Costa Campos. He lent me some books and some of his papers. I read them and I was very troubled, because there were several incompatible definitions and also incompatible with the theory of linear systems I knew. So, I decided to look for the derivative definition that had suitable properties for generalising to fractional orders the linear system concept and their usual characterizations namely the transfer function and impulse response. To do it, I decided to start from the Laplace transform domain and arrived to the Liouville derivative. With it I could introduce formally the fractional linear systems. I published the corresponding paper in 2000 and the main results are presented in Chap. 2. Meanwhile, I was far from the fractional community and the Fractional Calculus (FC) was a mere hobby. One day a colleague informed that there was at Porto another "crazy" person that worked also in FC: Prof. Tenreiro Machado. He is almost permanently involved in organization of conferences and challenged me to send papers. One of them treated the initial condition problem. Most people were using the Riemann–Liouville (RL) or Caputo (C) derivatives and indentified their "initial conditions" with those of the system. I disagreed and proposed an alternative that is also in Chap. 4. Meanwhile I received the visit of Blas Vinagre that showed be some of the current applications and talked also with Stephen Samko. The initial condition problem was not the unique reason why I refused to accept the RL and C formulations as correct derivative definitions. I did not accept the two-step procedure involved if their computations. It was like desiring to go from Lisbon to New York and instead of a flight Lisbon/New York did two flights as Lisbon/Paris and Paris/New York. Of course and by the same reasons I reject the sequential Miller-Ross derivative and also the Marchaud derivative. I use to say that I work under the minimum energy principle.

M. D. Ortigueira, *Fractional Calculus for Scientists and Engineers*, Lecture Notes in Electrical Engineering, 84, DOI: 10.1007/978-94-007-0747-4_1,

When looking for interesting papers on FC, I discovered the report "A long standing conjecture failed?" by Virginia Kiryakova, which challenged me to try to find a bridge between the Grünwald–Letnikov (GL) derivative and the others, namely Cauchy, RL, and C. I discovered that bridge on reading the paper "Differences of Fractional Order" by Diaz and Osler in the flight to Vienna where I was going to present my ideas on FC at that time. There I met Peter Kremple. It was the beginning of interesting discussions that unfortunately had to be finished. I met also Dumitru Baleano that was also a beginner in the fractional people community.

After the conference I discover how to pass from the GL to the generalised Cauchy derivative and from it I deduced a regularized derivative from where I arrived at the Liouville derivative. This approach showed clearly the difference between the forward and backward derivative, designations more interesting than the left and right then used. These results are presented in Chap. 3. It is important to remark that these formulations have an intrinsic property: the causality. While the forward derivatives are causal, the backward are anti-causal.

When applying the above derivatives to stochastic processes, mainly the fractional Brownian motion, I wondered about the possibility of defining a two-sided (centred) derivative. This was made necessary by the computation of the autocorrelation function of fractionally derivated white noise. I started from the symmetric integer order derivative and obtained two types of derivatives and showed their relations with the Riesz potentials. In the way I followed I obtained two Cauchy like derivatives. These are presented in Chap. 5.

May sound strange that I did not talk about Chap. 2. This has one reason: it was assumed that the GL was suitable for numerical implementation but not for theoretical results. With the help of Juan Trujillo I decided to use GL as a base for the foundation of FC. To do it we generalise the GL and studied its properties. Although the work is not finished (never will be!) in Chap. 2 it is shown that we can deduce all the other derivatives from GL. The main properties are deduced. One of them is the group property that states the exponent law is valid for any orders. This allows us to join in one formula the derivatives and primitives and solve one of the main subjects of discussions I had with P. Kremple: the integration constant and the complimentary polynomials. I concluded that they are not needed. We have a formulation valid for any order, positive or negative. In this case I suggest to use "anti-derivative". Its computation does not imply we have to insert any primitivation constant. This only appears when we are using the derivative rules to compute the primitives.

Some time ago I read something about the Quantum Derivative and thought it would be interesting to obtain a fractional formulation. I did it and proposed its use in formulating the theory of the linear scale invariant systems. This is discussed in Chap. 6. The presentation of The Fractional Quantum Derivative was done for the first time at the symposium on Applied Fractional Calculus held at Badajoz in 2007. I would like to say that I consider the symposia are much more interesting than other meetings since the availability people for discussions is greater. There, I met Richard Magin and Igor Podlubny and started our discussions.

Similarly the FSS in my Faculty was an important step in gaining insights into correct formulations and I hope that it will be the first of a series that will continue with the FSS2011.

In Chap. 7, I describe briefly some of interesting applications and give some insights into the future. I hope that much more people will join us in a near future with new applications and opening new horizons. The long memory character exhibited by the fractional systems makes them suitable for the description of a large set of Physical, Biological, Economical, … phenomena. Some of them may currently be described by nonlinear equations. We need more efficient tools for implementing and simulating fractional systems, mainly in discrete format. The current discrete approximations loose the long memory characteristic. We will not have to wait a long time for such realisations.

Chapter 2
The Causal Fractional Derivatives

2.1 Introduction

2.1.1 A Brief Historical Overview

The fractional calculus is a 300 years old mathematical discipline. In fact and some time after the publication of the studies on Differential Calculus, where he introduced the notation $\frac{d^n y}{dx^n}$, Leibnitz received a letter from Bernoulli putting him a question about the meaning of a non-integer derivative order. Also he received a similar enquiry from L'Hôpital: *What if n is ½? Leibnitz's replay was prophetic*: *It will lead to a paradox, a paradox from which one day useful consequences will be drawn, because there are no useless paradoxes.* It was the beginning of a discussion about the theme that involved other mathematicians like Euler and Fourier. Euler suggested in 1730 a generalisation of the rule used for computing the derivative of the power function. He used it to obtain derivatives of order 1/2. Nevertheless, we can say that the XVIII century was not proficuous in which concerns the development of Fractional Calculus. Only in the early XIX, interesting developments started being published. Laplace proposed an integral formulation (1812), but it was Lacroix who used for the first time the designation "derivative of arbitrary order" (1819). Using the gamma function he could define the fractional derivative of the power function, but did not go ahead. In 1822, Fourier presented the following generalization:

$$\frac{d^\nu f(t)}{dt^\nu} = \frac{1}{2\pi} \int_{-\infty}^{+\infty} f(\tau)\, d\tau \int_{-\infty}^{+\infty} u^\nu \cos(ut - u\tau + \nu\pi/2)\, du$$

and stated that it was valid for any ν, positive or negative. However, we can reference the beginning of the fractional calculus in the works of Liouville

M. D. Ortigueira, *Fractional Calculus for Scientists and Engineers*,
Lecture Notes in Electrical Engineering, 84, DOI: 10.1007/978-94-007-0747-4_2,

and Abel. Abel solved the integral equation that appears in the solution of the tautochrome problem[1]:

$$\int_a^t \frac{\varphi(\tau)}{(t-\tau)^\mu \, d\tau} = f(t), \quad t > a, \; 0 < \mu < 1$$

that represents an operation of fractional integration of order $1 - \mu$. However, it seems he was completely unaware of the fractional derivative or integral concept. Liouville did several attempts in 1832. In the first he took the exponentials as starting point for introducing the fractional derivative. With them he generalised the usual formula for the derivative of an exponential and applied it to the derivative computation of functions represented by series with exponentials (later called Dirichlet series). In another attempt he presented a formula for fractional integration similar to the above

$$D^{-p}\varphi(t) = \frac{1}{(-1)^p \Gamma(p)} \int_0^\infty \varphi(t+\tau)\tau^{p-1} \, d\tau, \quad -\infty < t < +\infty, \; \mathrm{Re}(p) > 0 \qquad (2.1)$$

where $\Gamma(p)$ is the gamma function. To this integral, frequently with the term $(-1)^p$ omitted, we give the name of Liouville's fractional integral. It is important to refer that Liouville was the first to consider the solution of differential equations. In other papers, Liouville went ahead with the development of ideas concerning this theme, having presented a generalization of the notion of incremental ratio to define a fractional derivative [1]. This idea was recovered, later, by Grünwald (1867) and Letnikov (1868). It is interesting to refer that Liouville was the first to note the difference between forward and backward derivatives (somehow different from the concepts of left and right). In a paper published in 1892 (after his death) Riemann reached to an expression similar to (2.1) for the fractional integral

$$D^{-\alpha}\varphi(t) = \frac{1}{\Gamma(\alpha)} \int_0^t \frac{\varphi(\tau)}{(t-\tau)^{1-\alpha}} \, d\tau, \quad t > 0 \qquad (2.2)$$

that, together with (2.2), became the more important basis for the fractional integration. It suits to refer that both Liouville and Riemann dealt with the called "complementary" functions that would appear when treating the differentiation of order α as an integration of order $-\alpha$. Holmgren (1865/66) and Letnikov (1868/74) discussed that problem when looking for the solution of differential equations, putting in a correct statement the fractional differentiation as inverse operation of the fractional integration. Besides, Holmgren gave a rigorous proof of Leibnitz'

[1] This refers to the problem of determining the shape of the curve such that the time of descent of a frictionless point mass sliding down the curve under the action of gravity is independent of the starting point.

rule for the fractional derivative of the product of two functions that was published before by Liouville, first, and Hargrave, later (1848). In the advent of XXth century, Hadamard proposed a fractional differentiation method by differentiating term by term the Taylor's series associated with the function. Weyl (1917) defined a fractional integration suitable to periodic functions, having used the integrals.

$$I_{+}^{\alpha}(t) = \frac{1}{\Gamma(\alpha)} \int_{-\infty}^{t} \frac{\varphi(\tau)}{(t-\tau)^{1-\alpha}} \mathrm{d}\tau, \quad -\infty < t < +\infty, \ 0 < \alpha < 1 \tag{2.3}$$

and

$$I_{-}^{\alpha}(t) = \frac{1}{\Gamma(\alpha)} \int_{t}^{\infty} \frac{\varphi(\tau)}{(\tau - t)^{1-\alpha}} \mathrm{d}\tau, \quad -\infty < t < +\infty, \ 0 < \alpha < 1 \tag{2.4}$$

particular cases of Liouville and Riemann ones but that have been, a basis for fractional integration in **R**. An interesting contribution to the fractional differentiation was given by Marchaud (1927) that presented a new formulation

$$D_{+}^{\alpha} f(t) = c \cdot \int_{0}^{\infty} \frac{\Delta_{\tau}^{k} f(t)}{\tau^{1+\alpha}} \mathrm{d}\tau, \quad \alpha > 0 \tag{2.5}$$

where $\Delta_{\tau}^{k} f(t)$ is the finite difference of order $k > \alpha$, $k = 1,2,3,\ldots$ and c a normalization constant. This definition coincides with[2]

$$D^{\alpha} f(t) = \frac{1}{\Gamma(\alpha - n)} \frac{\mathrm{d}^{n}}{\mathrm{d}t^{n}} \int_{-\infty}^{t} \frac{f(\tau)}{(t-\tau)^{\alpha - n + 1}} \mathrm{d}\tau, \quad n = \lfloor \alpha \rfloor + 1 \tag{2.6}$$

for enough "good" functions. It is important to remark that the construction (2.5) is advantageous relatively to (2.6): it can be applied to functions with "bad" behaviour at infinity, as being allowed to grow up as $t \to +\infty$.

A different approach was proposed by Heaviside (1892): the so-called operational calculus that was not readily accepted by the mathematicians till the works of Carson, Bromwich, and Doetsch that validated his procedure with help of the Laplace transform (LT).

Modernly, the unified formulation of integration and differentiation—called differintegration—based on Cauchy integral[3]

---

[2] $= \lfloor \alpha \rfloor$ means "integer part" of α.

[3] c is U-shaped integration path.

$$f^{(\alpha)}(z) = \frac{\Gamma(\alpha+1)}{2\pi j}\int_c \frac{f(\tau)}{(\tau - z)^{1+\alpha}} d\tau \tag{2.7}$$

gained great popularity. This approach can be referenced for the first time to Sonin (1869), but only with Laurent (1884) obtained a coherent formulation. In the eighties in the XX century this approach evolved with the works of several people as Nishimoto (published a sequence of four books), Campos, Srivastava, Kalla, Riesz, Osler, etc. The book of Samko et al. [1] was the culminate of several step development. However, the main part of the book is devoted, not to the Cauchy integral formulation, but to the so-called Riemann–Liouville derivative. This formulation appeared first in a paper by Sonin (1869) and joins the Riemann and Liouville integral formulations together with the integer order derivative. Essentially it is a multi-step procedure that does an integer order derivative after a fractional integration. Caputo in the sixties inverted the procedure: one starts by an integer order derivative and afterwards does a fractional integration. We must refer also a different form of fractional differentiation that was introduced by Riesz: the so-called Riesz and Riesz-Feller potentials.

Since the beginning of the nineties of XXth century, the Fractional Calculus attracted the attention of an increasing number of mathematicians, physicians, and engineers that have been supporting its development and originating several new formulations and mainly using it to explain some natural and engineering phenomena and also using it to develop new engineering applications.

2.1.2 Current Formulations

In Tables 2.1 and 2.2, the most known definitions of Fractional Integrals and Derivatives are presented.

As seen, there are clear differences among some kinds of definitions. On the other hand, there are definitions that impose causality and the relation

$$\mathrm{LT}[D^\alpha f(t)] = s^\alpha F(s), \quad \alpha \in R \tag{2.8}$$

is not always valid. However, the major inconvenient of most definitions is in the fact of incorporating properties of the signal. Although we can talk about derivatives or integrals of functions defined on a given sub-interval in R, but we do not find correct to incorporate that property in the definition. This means that only the definitions with R as domain may be valid definitions. This assumption brings an important consequence: the integral and derivative are inverse operations and commute:

$$D^\alpha\{D^\beta\} = D^{\alpha+\beta} = D^\beta\{D^\alpha\}, \quad \alpha, \beta \in R$$

Later we will return to this subject.

Table 2.1 Fractional integral definitions ($\alpha > 0$)

Designation	Definition
Liouville integral	$D^{-\alpha}\varphi(t) = \frac{1}{(-1)^{\alpha}\Gamma(\alpha)}\int_0^{+\infty}\varphi(t+\tau)\tau^{\alpha-1}\,\mathrm{d}\tau$
Riemann integral	$D^{-\alpha}\varphi(t) = \frac{1}{\Gamma(\alpha)}\int_0^{t}\varphi(\tau)\cdot(t-\tau)^{\alpha-1}\,\mathrm{d}\tau, \quad t>0$
Hadamard integral	$D^{-\alpha}(t) = \frac{t^{\alpha}}{\Gamma(\alpha)}\int_0^{1}\varphi(t\tau)\cdot(1-\tau)^{\alpha-1}\,\mathrm{d}\tau, \quad t>0$
Left side Riemann–Liouville integral	$D^{-\alpha}\varphi(t) = \frac{1}{\Gamma(\alpha)}\int_a^{t}\varphi(\tau)\cdot(t-\tau)^{\alpha-1}\,\mathrm{d}\tau, \quad t>a$
Right side Riemann–Liouville integral	$D^{-\alpha}\varphi(t) = \frac{1}{\Gamma(\alpha)}\int_t^{b}\varphi(\tau)\cdot(\tau-t)^{\alpha-1}\,\mathrm{d}\tau, \quad t<b$
Left side Weyl integral	$D^{-\alpha}\varphi(t) = \frac{1}{\Gamma(\alpha)}\int_{-\infty}^{t}\varphi(\tau)\cdot(t-\tau)^{\alpha-1}\,\mathrm{d}\tau$
Right side Weyl integral	$D^{-\alpha}\varphi(t) = \frac{1}{\Gamma(\alpha)}\int_t^{+\infty}\varphi(\tau)\cdot(\tau-t)^{\alpha-1}\,\mathrm{d}\tau$

Table 2.2 Fractional derivative definitions ($\alpha > 0$)

Designation	Definition
Left side Riemann–Liouville derivative	$D^{\alpha}\varphi(t) = \frac{1}{\Gamma(n-\alpha)}\frac{\mathrm{d}n}{\mathrm{d}t^n}\int_a^{t}\varphi(\tau)\cdot(t-\tau)^{\alpha-n-1}\,\mathrm{d}\tau$
Right side Riemann–Liouville derivative	$D^{\alpha}\varphi(t) = \frac{(-1)^n}{\Gamma(n-\alpha)}\frac{\mathrm{d}^n}{\mathrm{d}t^n}\int_t^{b}\varphi(\tau)\cdot(\tau-t)^{\alpha-n-1}\,\mathrm{d}\tau$
Left side Caputo derivative	$D^{\alpha}\varphi(t) = \frac{1}{\Gamma(-\nu)}\left[\int_0^{t}\varphi^{(n)}(\tau)\cdot(t-\tau)^{\nu-1}\,\mathrm{d}\tau\right], \quad t>0$
Right side Caputo derivative	$D^{\alpha}\varphi(t) = \frac{1}{\Gamma(-\nu)}\left[\int_t^{+\infty}\varphi^{(n)}(\tau)\cdot(\tau-t)^{\nu-1}\,\mathrm{d}\tau\right]$
Marchaud derivative	$D^{\alpha}_{+}f(t) = c\cdot\int_0^{\infty}\frac{\Delta^k_{\tau}f(t)}{\tau^{1+\alpha}}\,\mathrm{d}\tau, \quad \alpha>0$
Generalised function	$D^{\alpha}\varphi(t) = \frac{1}{\Gamma(-\alpha)}\int_{-\infty}^{t}\varphi(\tau)\cdot(t-\tau)^{-\alpha-1}\,\mathrm{d}\tau$
Left Grünwald–Letnikov	$D^{\alpha}_{-}f(t) = \lim_{h\to 0+}\frac{\sum_{k=0}^{\infty}(-1)^k\binom{\alpha}{k}f(t-kh)}{h^{\alpha}}$
Right Grünwald–Letnikov	$D^{\alpha}_{+}f(t) = \lim_{h\to 0+}\frac{\sum_{k=0}^{\infty}(-1)^k\binom{\alpha}{k}f(t+kh)}{h^{\alpha}}$

2.1.3 A Signal Processing Point of View

In recent years fractional calculus has been rediscovered by scientists and engineers and applied in an increasing number of fields, namely in the areas of electromagnetism, control engineering, and signal processing. The increase in the number of physical and engineering processes that are best described by fractional differential equations has motivated out its study. This led to an enrichment of Fractional Calculus with new approaches that however brought contributions to a somehow chaotic state of the art. As seen, there are several definitions that lead to different results, making difficult the establishment of a systematic theory of fractional linear systems in agreement with the current practice. Although from a purely mathematical point of view it is legitimate to accept and even use one or all, from the point of view of applications, the situation is different. We should accept only the definitions that might lead to a fractional systems theory coherent with the usual practice and accepted notions and concepts such as the impulse response and transfer function. The use of the Grünwald–Letnikov forward and backward derivatives lead to a correct generalization of the current linear systems theory. Moreover this choice is motivated by other reasons:

- It does not need superfluous derivative computations.
- It does not insert unwanted initial conditions.
- It is more flexible.
- It allows sequential computations.
- It leads to the other definitions.

On the other hand, it does not assume any bound on the domain of the signals to be used. In general we will assume it is the real line. If we want to use any bounded interval we will use the Heaviside unit step function. This has as consequence that we will use the bilateral (two-sided) Laplace transform. We will not use the one-sided LT, for several reasons:

- It forces us to use only causal signals.
- Some of its properties lose symmetry, e.g. the translation and the derivation/integration properties.
- It does not treat easily the case of impulses located at $t = 0$ [2].
- In the fractional case, it imposes on us the same set of initial conditions as the Riemann–Liouville case that can be a constraint.

2.1.4 Overview

This chapter has three main parts corresponding to the fractional derivatives definitions, their properties and generalisations. We present a general formula and the forward and backward derivatives as special cases valid for real functions. We treat the case of functions with Laplace transform and obtain integral formulae named as Liouville differintegrations, since they were proposed first by Liouville. These are suitable for fractional linear systems studies, since they allow a generalisation of known concepts without meaningful changes. We will show that these derivatives impose causality: one is causal and the other anti-causal. For the general derivative we will prove its semi-group properties and deduce some other interesting features. We will show that it is compatible with classic derivative that appears here as a special case. We will compute the derivatives of some useful functions. In particular, we obtain derivatives of exponentials, causal exponentials, causal powers and logarithms.

2.1.5 Cautions

We deal with a multivalued expression z^{α}. As is well known, to define a function we have to fix a branch cut line and choose a branch (Riemann surface). It is a common procedure to choose the negative real half-axis as branch cut line. In what follows we will assume that we adopt the principal branch and assume that the obtained function is continuous above the branch cut line. With this, we will write $(-1)^{\alpha} = e^{j\alpha\pi}$.

In the following and otherwise stated, we will assume to be in the context of the generalised functions (distributions). We always assume that they are either of exponential order or tempered distributions.

Unless stated, our domain of work will be the entire **R**, not **R**$^{+}$, or, when stated **C**.

As referred before, "our" Laplace transform (LT) will be the two-sided Laplace transform.

2.2 From the Classic Derivative to the Fractional

2.2.1 On the Grünwald–Letnikov Derivative

In the prehistory of Fractional Calculus, Liouville (1832) was the first to look for a definition of fractional derivative through the generalization of the incremental ratio used for integer order derivatives to the fractional case [1, 3]. However, he did not go on with this idea. Greer (1859) treated the order ½ case. Grünwald (1867) and mainly Letnikov (1868,1872) studied the fractional derivative obtained by the referred generalization and studied its properties. Here we will present a more general vision of the incremental ratio derivative and deduce its properties. The most important is the semi-group property that created great difficulties in the past. In fact, this seems to have been considered first by Peacock under the "principle of the permanence of equivalent forms". However, he did not convince anybody. The same happened to Kelland (1846) and later to Heaviside in the nineties in the XIX century. However, Heaviside got interesting results with his operational calculus that contributed to be accepted in several scientific domains. But in Fractional Calculus, the group property has only been accepted in the integral case. We will show that it is valid in the general case and maintained in the generalized functions case. The main point is in the use of the same formula for both derivative case (positive orders) and integral case (negative orders). In this case, we do not have to care neither about any integration constant, nor on initial conditions.

2.2.2 Difference Definitions

Let $f(z)$ be a complex variable function and introduce Δ_d and Δ_r as finite "direct" and "reverse" differences defined by:

$$\Delta_d f(z) = f(z) - f(z-h) \tag{2.9}$$

and

$$\Delta_r f(z) = f(z+h) - f(z) \tag{2.10}$$

with $h \in C$ and, for reasons that will be apparent later, we assume that $\mathrm{Re}(h) > 0$ or $\mathrm{Re}(h) = 0$ with $\mathrm{Im}(h) > 0$. The repeated use of the above definitions lead to

$$\Delta_d^N f(z) = \sum_{k=0}^{N} (-1)^k \binom{N}{k} f(z-kh) \tag{2.11}$$

and

$$\Delta_r^N f(z) = (-1)^N \sum_{k=0}^{N} (-1)^k \binom{N}{k} f(z+kh) \tag{2.12}$$

where $\binom{N}{k}$ are the binomial coefficients. These definitions are readily extended to the fractional order case [4]:

$$\Delta_d^{\alpha} f(z) = \sum_{k=0}^{\infty} (-1)^k \binom{\alpha}{k} f(z-kh) \tag{2.13}$$

and

$$\Delta_r^{\alpha} f(z) = (-1)^{\alpha} \sum_{k=0}^{\infty} (-1)^k \binom{\alpha}{k} f(z+kh) \tag{2.14}$$

where we assume that $\alpha \in R$. This formulation remains valid in the negative integer case. Let $\alpha = -N$ (N a positive integer). As it is well known from the Z Transform theory, the following relation holds, if $k \geq 0$

$$ZT[(n+1)_k u(n)] = \frac{k!}{(1-Z^{-1})^{k+1}} \quad \text{for } |q| > 1 \tag{2.15}$$

where $u(n)$ is the discrete time unit step:

$$u(n) = \begin{cases} 1 & n \geq 0 \\ 0 & n < 0 \end{cases}$$

Introducing the Pochhammer symbol,

$$(a)_k = a(a+1)(a+2)\cdots(a+k-1)$$

and putting $k = N - 1$, we obtain easily:

$$(1-q^{-1})^{-N} = \sum_{n=0}^{\infty} \frac{(n+1)_{N-1}}{(N-1)!} q^{-n} \quad \text{for } |q| > 1 \tag{2.16}$$

Interpreting q^{-1} as a delay as it is commonly done in Digital Signal Processing, (2.16) leads to

$$\Delta_d^{-N} f(z) = \sum_{n=0}^{\infty} \frac{(n+1)_{N-1}}{(N-1)!} f(z-nh) \tag{2.17}$$

For the reverse case, we have:

$$\Delta_r^{-N} f(z) = (z-1)^{-N} f(z) = (-1)^N \sum_{n=0}^{\infty} \frac{(n+1)_{N-1}}{(N-1)!} f(z+nh) \tag{2.18}$$

As

$$(n+1)_{N-1} = \frac{(n+N-1)!}{n!} = \frac{(N-1)!(N)_n}{n!} \tag{2.19}$$

and

$$\frac{(-a)_n}{n!} = (-1)^n \binom{a}{n} \tag{2.20}$$

we have:

$$(n+1)_{N-1} = (N-1)!(-1)^n \binom{-N}{n} \tag{2.21}$$

So, we can write:

$$\Delta_d^{-N} f(z) = \sum_{n=0}^{\infty} (-1)^n \binom{-N}{-n} f(z-nh) \tag{2.22}$$

For the anti-causal case, we have:

$$\Delta_r^{-N} f(z) = (-1)^N \sum_{n=0}^{\infty} (-1)^n \binom{-N}{n} f(z+nh) \tag{2.23}$$

As it can be seen, these expressions are the ones we obtain by putting $\alpha = -N$ into (2.13) and (2.14) that emerge here as representations for the differences of any order.

2.2.3 Integer Order Derivatives

The normal way of introducing the derivative of a continuous function is through the limits of the incremental ratio:

$$f_d^{(1)}(z) = \lim_{h\to 0} \frac{f(z) - f(z-h)}{h} \tag{2.24}$$

and

$$f_r^{(1)}(z) = \lim_{h\to 0} \frac{f(z+h) - f(z)}{h} \tag{2.25}$$

These incremental ratios are very important in continuous to discrete conversion of systems defined by differential equations (linear or nonlinear). It is also well known that the first is better than the second because of stability matters. The use of the LT into the above definitions allows us to obtain the transfer functions of the differentiators that are equal both in analytical expression and domain (the whole complex plane):

$$s = \lim_{h\to 0} \frac{(1 - e^{-sh})}{h} = \lim_{h\to 0} \frac{(e^{sh} - 1)}{h} \tag{2.26}$$

This is the reason why the above derivatives give the same result, whenever they exist and $f(z)$ is a continuous function. We must stress also that $h \in R$. Later we will see that in the general case h is a value on a half straight line in the complex plane.

In most books on Mathematical Analysis we are told that to compute the high order derivatives we must proceed sequentially by repeating the application of formulae (2.24) or (2.25). This means that, if we want to compute the fourth order derivative, we have to compute $f'(z)$, $f''(z)$, and $f^{(3)}(z)$. However, we have an alternative as we will see next. Assume that we want to compute the second order derivative from the first. We have

$$\begin{aligned} f_d^{(2)}(z) = \lim_{h\to 0} \frac{f^{(1)}(z) - f^{(1)}(z-h)}{h} &= \lim_{h\to 0} \frac{\lim_{h\to 0} \frac{f(z)-f(z-h)}{h} - \lim_{h\to 0} \frac{f(z-h)-f(z-2h)}{h}}{h} \\ &= \lim_{h\to 0} \frac{\lim_{h\to 0} \frac{f(z)-2f(z-h)+f(z-2h)}{h}}{h} \\ &= \lim_{h\to 0} \frac{f(z) - 2f(z-h) + f(z-2h)}{h^2} \end{aligned} \tag{2.27}$$

As seen, we obtained an expression that allows us to obtain the second order derivative directly from the function. It is not a difficult task to repeat the procedure for successively increasing orders to obtain a general expression:

$$f_d^{(n)}(z) = \lim_{h\to 0} \frac{\sum_{k=0}^{n} (-1)\binom{n}{k} f(z-kh)}{h^n} \tag{2.28}$$

that allows us to obtain the *n*th order derivative directly without "passing" by the intermediate derivatives. To see that this is correct, assume that we want to compute the $(n+1)$th order derivative of $f(t)$ from the first order derivative of (2.28). We have:

$$\begin{aligned}
f_d^{(n+1)}(z) &= \lim_{h\to 0} \frac{\lim_{h\to 0}\frac{\sum_{k=0}^{n}(-1)^k\binom{n}{k}f(z-kh)}{h^n} - \lim_{h\to 0}\frac{\sum_{k=0}^{n}(-1)^k\binom{n}{k}f(z-kh-h)}{h^n}}{h} \\
&= \lim_{h\to 0} \frac{\lim_{h\to 0}\frac{\sum_{k=0}^{n}(-1)^k\binom{n}{k}f(z-kh)-\sum_{k=0}^{n}(-1)^k\binom{n}{k}f(z-kh-h)}{h^n}}{h} \\
&= \frac{\sum_{k=0}^{n}(-1)^k\binom{n}{k}f(z-kh) - \sum_{k=0}^{n}(-1)^k\binom{n}{k}f(z-kh-h)}{h^{n+1}} \\
&= \frac{\sum_{k=0}^{n}(-1)^k\binom{n}{k}f(z-kh) + \sum_{k=1}^{n+1}(-1)^k\binom{n}{k-1}f(z-kh)}{h^{n+1}} \\
&= \frac{\sum_{k=0}^{n+1}(-1)^k\left[\binom{n}{k}+\binom{n}{k-1}\right]f(z-kh)}{h^{n+1}}
\end{aligned} \tag{2.29}$$

As, $(-1)! = \infty$ and $0! = 1$, we conclude easily that

$$\binom{n}{k} + \binom{n}{k-1} = \binom{n+1}{k}$$

which, together with (2.29), confirms the validity of relation (2.28). Proceeding similarly with (2.25) we obtain

$$f_r^{(n)}(z) = (-1)^n \lim_{h\to 0} \frac{\sum_{k=0}^{n}(-1)^k\binom{n}{k}f(z+kh)}{h^n} \tag{2.30}$$

Expressions (2.28) and (2.30) allow us to do a direct computation of the *n*th order derivative of a given function. Considering (2.11) and (2.12) we see that the above derivatives are limits of incremental ratio. As before it is a simple task to use the LT to obtain the transfer function of the differentiator

$$H(s) = s^n \tag{2.31}$$

valid for $s \in C$. From this result we conclude that the differentiator is a linear system that does not impose causality. It is what we may nominate *acausal* system. This is a very important subject that will be invalid when generalising the derivative concept.

2.3 Definition of Fractional Derivative

To generalize the known notion of fractional derivatives we start from the above derivatives to introduce the general formulation of the incremental ratio valid for any order, real or complex, obtained from the fractional order differences (2.13) and (2.14).

Definition 2.1 We define fractional derivative of $f(z)$ by the limit of the fractional incremental ratio

$$D_{\theta}^{\alpha}f(z) = \mathrm{e}^{-j\theta\alpha} \lim_{|h|\to 0} \frac{\sum_{k=0}^{n} (-1)^k \binom{\alpha}{k} f(z-kh)}{|h|^{\alpha}} \tag{2.32}$$

where $h = |h|\mathrm{e}^{j\theta}$ is a complex number, with $\theta \in (-\pi, \pi]$. This derivative is a general incremental ratio based derivative that expands to the whole complex plane the classic derivatives and the Grünwald–Letnikov fractional derivatives. We will retain this name and refer as GL derivative in the following.

To understand and give an interpretation to the above formula, assume that z is a time and that h is real, $\theta = 0$ or $\theta = \pi$. If $\theta = 0$, only the present and past values are being used, while, if $\theta = \pi$, only the present and future values are used. This means that if we look at (2.32) as a linear system, the first case is causal, while the second is anti-causal[4] [5, 6].

In general, if $\theta = 0$, we call (2.32) the *forward* Grünwald–Letnikov[5] derivative.

$$D_{f}^{\alpha}f(z) = \lim_{h\to 0+} \frac{\sum_{k=0}^{\infty} (-1)^k \binom{\alpha}{k} f(z-kh)}{h^{\alpha}} \tag{2.33}$$

If $\theta = \pi$, we put $h = -|h|$ to obtain the *backward* Grünwald–Letnikov derivative.

$$D_{b}^{\alpha}f(z) = \lim_{h\to 0+} \mathrm{e}^{-j\pi\alpha} \frac{\sum_{k=0}^{\infty} (-1)^k \binom{\alpha}{k} f(z+kh)}{h^{\alpha}} \tag{2.34}$$

The exponential factor in this formula makes it different from the so-called right GL derivative found in current literature.

4 We will return to this matter later.

5 The terms forward and backward are used here in agreement to the way the time flows, from past to future or the reverse.

2.4 Existence

It is not a simple task to formulate the weakest conditions that ensure the existence of the fractional derivatives (2.32), (2.33) and (2.34), although we can give some necessary conditions for their existence. To study the existence conditions for the fractional derivatives we must care about the behaviour of the function along the half straight-line $z \pm nh$ with $n \in Z^+$. If the function is zero for $\mathrm{Re}(z) < a \in R$ (resp. $\mathrm{Re}(z) > a$) the forward (backward) derivative exists at every finite point of $f(z)$. In the general case, we must have in mind the behavior of the binomial coefficients. They verify

$$|\binom{\alpha}{k}| \leq \frac{A}{k^{\alpha+1}} \tag{2.35}$$

meaning that $f(z).\frac{A}{k^{\alpha+1}}$ must decrease, at least as $\frac{A}{k^{|\alpha|+1}}$ when k goes to infinite.

For instance, we are going to consider the forward case. If $\alpha > 0$, it is enough to ensure that $f(z)$ is bounded in the left half plane; but if $\alpha < 0$, $f(z)$ must decrease to zero to obtain a convergent series. This suggests that the behaviour for $\mathrm{Re}(z) < 0$ or $\mathrm{Re}(z) > 0$ should be adopted for defining right and left functions. We say that $f(z)$ is a right [left] function if $f(-\infty) = 0$ $[f(+\infty) = 0]$. In particular, they should be used for the functions such that $f(z) = 0$ for $\mathrm{Re}(z) < 0$ and $f(z) = 0$ for $\mathrm{Re}(z) > 0$, respectively.[6] This is very interesting, since we conclude that the existence of the fractional derivative depends only on what happens in one half complex plane, left or right. Consider $f(z) = z^\beta$, with $\beta \in R$ with a suitable branch cut line. If $\beta > \alpha$, we conclude immediately that $D^\alpha[z^\beta]$ defined for every $z \in C$ does not exist, unless α is a positive integer, because the summation in (2.32) is divergent.

2.5 Properties

We are going to present the main properties of the above presented derivative.

Linearity The linearity property of the fractional derivative is evident from the above formulae. In fact, we have

$$D^\alpha_\theta[f(z) + g(z)] = D^\alpha_\theta f(z) + D^\alpha_\theta g(z) \tag{2.36}$$

Causality The causality property was already referred above and can also be obtained easily. We must be ware that it only makes sense, if we are using the forward or backward derivatives and that $t = z \in R$. Assume that $f(t) = 0$, for

[6] By breach of language we call them causal and anti-causal functions borrowing the system terminology.

$t < 0$. We conclude immediately that $D_{\theta}^{\alpha}f(t) = 0$ for $t < 0$. For the anti-causal case, the situation is similar.

Scale change Let $f(z) = g(az)$, where a is a constant. Let $h = |h|e^{j\theta}$ and $a = |a|e^{j\varphi}$. From (2.32), we have:

$$\begin{aligned} D_{\theta}^{\alpha}g(az) &= \lim_{h\to 0}\frac{\sum_{k=0}^{\infty}(-1)^k\binom{\alpha}{k}g(az-kah)}{h^{\alpha}} \\ &= a^{\alpha}\lim_{h\to 0}\frac{\sum_{k=0}^{\infty}(-1)^k\binom{\alpha}{k}g(az-kah)}{(ah)^{\alpha}} \\ &= a^{\alpha}D_{\theta+\varphi}^{\alpha}g(\tau)|_{\tau=az} \end{aligned} \tag{2.37}$$

Time reversal If z is a time and $f(z) = g(-z)$, we obtain from the property we just deduced that:

$$\begin{aligned} D_{\theta}^{\alpha}g(-z) &= (-1)^{\alpha}\lim_{h\to 0}\frac{\sum_{k=0}^{\infty}(-1)^k\binom{\alpha}{k}g(-z+kah)}{(-h)^{\alpha}} \\ &= (-1)^{\alpha}D_{\theta}^{\alpha}g(\tau)|_{\tau=-z} \end{aligned} \tag{2.38}$$

in agreement with (2.33) and (2.34). This means that the time reversal converts the forward derivative into the backward and vice versa.

Shift invariance The derivative operator is shift invariant:

$$D_{\theta}^{\alpha}f(z-a) = D_{\theta}^{\alpha}g(\tau)|_{\tau=z-a} \tag{2.39}$$

as it can be easily verified.

Derivative of a product We are going to compute the derivative of the product of two functions—$f(t) = \varphi(t)\cdot\psi(t)$—assumed to be defined for $t \in R$, by simplicity, although the result we will obtain is valid for $t \in C$, excepting over an eventual branch cut line. Assume that one of them is analytic in a given region. From (2.32) and working with increments, we can write

$$\Delta^{\alpha}f(z) = \frac{\sum_{k=0}^{\infty}(-1)^k\binom{\alpha}{k}\varphi(z-kh)\psi(z-kh)}{h^{\alpha}} \tag{2.40}$$

But, as

$$\Delta^{N}f(z) = \sum_{k=0}^{N}(-1)^k\binom{N}{k}f(z-kh) \tag{2.41}$$

we can obtain

$$f(z-kh) = \sum_{i=0}^{k}(-1)^i\binom{k}{i}\Delta^{i}f(z) \tag{2.42}$$

that inserted in (2.40) leads to

$$\Delta^{\alpha} f(z) = \frac{\sum_{k=0}^{\infty} (-1)^k \binom{\alpha}{k} \varphi(z-kh) \sum_{i=0}^{k} (-1)^i \binom{k}{i} \Delta^i \psi(z)}{h^{\alpha}} \tag{2.43}$$

that can be transformed into:

$$\Delta^{\alpha} f(z) = \frac{\sum_{i=0}^{\infty} (-1)^i \Delta^i \psi(z) \sum_{k=i}^{\infty} (-1)^k \binom{k}{i} \binom{\alpha}{k} \varphi(z-kh)}{h^{\alpha}} \tag{2.44}$$

But

$$(-1)^{k+i} \binom{k+i}{i} \binom{\alpha}{k+i} = \frac{(-\alpha)_i}{i!} \frac{(-\alpha+i)_k}{k!} = \frac{(-\alpha)_i}{i!} (-1)^k \binom{\alpha-i}{k}$$

that substituted into the above relation gives

$$\Delta^{\alpha} f(z) = \frac{\sum_{i=0}^{\infty} \binom{\alpha}{i} \Delta^i \psi(z) \sum_{k=0}^{\infty} (-1)^k \binom{\alpha-i}{k} \varphi(z-kh-ih)}{h^{\alpha}} \tag{2.45}$$

and

$$\Delta^{\alpha} f(z) = \sum_{i=0}^{\infty} \frac{\binom{\alpha}{i} \Delta^i \psi(z)}{h^i} \frac{\sum_{k=0}^{\infty} (-1)^k \binom{\alpha-i}{k} \varphi(z-kh-ih)}{h^{\alpha-i}} \tag{2.46}$$

Computing the limit as $h \to 0$, we obtain the derivative of the product:

$$D_{\theta}^{\alpha}[\varphi(t)\psi(t)] = \sum_{n=0}^{\infty} \binom{\alpha}{n} \varphi^{(n)}(t) \psi^{(\alpha-n)}(t) \tag{2.47}$$

that is the generalized Leibniz rule. This formula was obtained first by Liouville [7]. We must realize that the above formula is commutative if both functions are analytic. If only one of them is analytic, it is not commutative. We must remark that the noncommutativity of this rule seems natural, since we only require analyticity to one function. It is a situation very similar to the one we find when defining the product of generalized functions and its derivatives.

The deduction of (2.47) we presented here differs from others presented in literature [1]. As it is clear when $\alpha = N \in Z^+$ we obtain the classic Leibniz rule. When $\alpha = -1$, we obtain a very interesting formula for computing the primitive of the product of two functions, generalizing the partial primitivation:

$$D^{-1}[\varphi(t)\psi(t)] = \sum_{n=0}^{\infty} (-1)^n \varphi^{(n)}(t) \psi^{(-n-1)}(t) \tag{2.48}$$

This formula can be useful in computing the LT of the Fourier transform (FT). We only have to choose $\varphi(t)$ or $\psi(t)$ equal to e^{-st} in the LT case and equal to $e^{-j\omega t}$ in the FT case.

To exemplify the use of this formula let $g(t) = \frac{t^n}{n!} \cdot e^{at}$. Put $\varphi(t) = \frac{t^n}{n!}$ and $\psi(t) = e^{at}$. As $\varphi^{(k)}(t) = \frac{t^{n-k}}{(n-k)!}$, while $k \leq n$ and $\psi^{(-k-1)}(t) = a^{-k-1}e^{at}$. Then:

$$D^{-1}g(t) = e^{at}\sum_{k=0}^{n}(-1)^k\frac{t^{n-k}}{(n-k)!}a^{-k-1}$$

From this formula, we can compute the LT of $\frac{t^n}{n!}u(t)$. If we put $\varphi(t) = f(t)$ and $\psi(t) = 1$, $t \in R$, we have:

$$D^{-1}[f(t)] = \sum_{n=0}^{\infty}(-1)^n f^{(n)}(t)\frac{t^{n+1}}{(n+1)!} \tag{2.49}$$

similar to the McLaurin formula.

Integration by parts The so-called integration by parts relates both causal and anti-causal derivatives and can be stated as:

$$\int_{-\infty}^{+\infty} g(t)D_f^{\alpha}f(t)\,dt = (-1)^{\alpha}\int_{-\infty}^{+\infty} f(t)D_b^{\alpha}g(t)\,dt \tag{2.50}$$

where we assume that both integrals exist. To obtain this formula, we only have to use (2.33) inside the integral and perform a variable change

$$\int_{-\infty}^{+\infty}\lim_{h\to 0+}\frac{\sum_{k=0}^{\infty}(-1)^k\binom{\alpha}{k}f(t-kh)g(t)}{h^{\alpha}}\,dt$$

$$= \int_{-\infty}^{+\infty}\lim_{h\to 0+}\frac{\sum_{k=0}^{\infty}(-1)^k\binom{\alpha}{k}g(t+kh)f(t)}{h^{\alpha}}\,dt$$

leading immediately to (2.50) if we use (2.32). This result is slightly different from the one find in current literature due to our definition of backward derivative.

2.6 Group Structure of the Fractional Derivative

2.6.1 *Additivity and Commutativity of the Orders*

Additivity We are going to apply (2.32) twice for two orders. We have

$$D_{\theta}^{\alpha}[D_{\theta}^{\beta}f(t)] = D_{\theta}^{\beta}[D_{\theta}^{\alpha}f(t)] = D_{\theta}^{\alpha+\beta}f(t) \tag{2.51}$$

To prove this statement we start from (2.32) and write:

$$D_\theta^\alpha\left[D_\theta^\beta f(t)\right] = \lim_{h\to 0}\frac{\sum_{k=0}^{\infty}\binom{\alpha}{k}(-1)^k\left[\sum_{n=0}^{\infty}\binom{\beta}{n}(-1)^n f[t-(k+n)h]\right]}{h^\alpha h^\beta}$$
$$= \lim_{h\to 0}\frac{\sum_{n=0}^{\infty}\binom{\beta}{n}(-1)^n\left[\sum_{k=0}^{\infty}\binom{\alpha}{k}(-1)^k f[t-(k+n)h]\right]}{h^\alpha h^\beta}$$

for any $\alpha, \beta \in R$, or even $\in C$. With a change in the summation, we obtain:

$$D_\theta^\alpha\left[D_\theta^\alpha f(t)\right] = \lim_{h\to 0}\frac{\sum_{m=0}^{\infty}\left[\sum_{n=0}^{\infty}\binom{\alpha}{m-n}\binom{\beta}{n}\right](-1)^m f[t-mh]}{h^{\alpha+\beta}}$$

As Samko et al. [1]

$$\sum_0^m \binom{\beta}{m-n}\binom{\beta}{n} = \binom{\alpha+\beta}{m}, \tag{2.52}$$

$$D_\theta^\alpha\left[D_\theta^\beta f(t)\right] = \lim_{h\to 0}\frac{\sum_{m=0}^{\infty}\binom{\alpha+\beta}{m}(-1)^m f[t-mh]}{h^{\alpha+\beta}} = D^{\alpha+\beta}f(t)$$

Associativity This property comes easily from the above results. In fact, it is easy to show that

$$D_\theta^\gamma\left[D_\theta^{\alpha+\beta}f(t)\right] = D_\theta^{\gamma+\alpha+\beta}f(t) = D_\theta^{\alpha+\beta+\gamma}f(t) = D_\theta^\alpha\left[D_\theta^{\beta+\gamma}f(t)\right] \tag{2.53}$$

Neutral element
If we put $\beta = -\alpha$ in (2.51) we obtain:

$$D_\theta^\alpha\left[D_\theta^{-\alpha}f(t)\right] = D_\theta^0 f(t) = f(t) \tag{2.54}$$

or using it again

$$D_\theta^{-\alpha}\left[D_\theta^\alpha f(t)\right] = D_\theta^0 f(t) = f(t) \tag{2.55}$$

This is very important because it states the existence of inverse.

Inverse element From the last result we conclude that there is always an inverse element: for every α order derivative, there is always a $-\alpha$ order derivative. This seems to be contradictory with our knowledge from the classic calculus where the Nth order derivative has N primitives. To understand the situation we must refer that the inverse is given by (2.32) and it does not give any primitivation constant. This forces us to be consistent and careful with the used language. So, when α is positive we will speak of derivative. When α is negative, we will use the term anti-derivative (not primitive or integral). This clarifies the situation and solves the problem created by Liouville and Riemann when they introduced the complimentary polynomials and shows how to obtain the "proper primitives" of Krempl [8]. The problem can be well understood if we realize that currently we do not have a direct way of computing integrals: we reverse the rules of differentiation.

This fact leads us to include the primitivation constants that are inserted artificially. In fact, having autonomous rules for primitivation that do not pass by the rules of differentiation the primitivation constant would not appear. It is what happen when we use the GL derivative (2.32) with negative orders. Also, when performing a numerical computation we do not have to care about such constant. Besides, we must refer that this does not have any relation with the initial conditions that appear in the solution of differential equations. Later we will return to this problem.

2.7 Simple Examples

2.7.1 The Exponential

Let us apply the above definitions to the function $f(z) = \mathrm{e}^{sz}$. The convergence of (2.33) is dependent of s and of h. Let $h > 0$, the series in (2.33) becomes

$$\mathrm{e}^{sz} \sum_{k=0}^{\infty} (-1)^k \binom{\alpha}{k} \mathrm{e}^{-ksh}$$

As it is well known, the binomial series

$$\sum_{k=0}^{\infty} (-1)^k \binom{\alpha}{k} \mathrm{e}^{-ksh}$$

is convergent to the main branch of

$$g(s) = (1 - \mathrm{e}^{-sh})^{\alpha}$$

provided that $|\mathrm{e}^{-sh}| < 1$, that is if $\mathrm{Re}(s) > 0$. This means that the branch cut line of $g(s)$ must be in the left hand half complex plane. Then

$$D_f^{\alpha} f(z) = \lim_{h \to 0+} \frac{(1 - \mathrm{e}^{-sh})}{h^{\alpha}} \mathrm{e}^{sz} = \lim_{h \to 0+} \left(\frac{1 - \mathrm{e}^{-sh}}{h} \right)^{\alpha} \mathrm{e}^{sz} = |s|^{\alpha} \mathrm{e}^{j\theta\alpha} \mathrm{e}^{sz} \tag{2.56}$$

iff $\theta \in (-\pi/2, \pi/2)$ which corresponds to be working with the principal branch of the power function, $(\cdot)^{\alpha}$, and assuming a branch cut line in the left hand complex half plane.

Now, consider the series in (2.34) with $f(z) = \mathrm{e}^{sz}$. Proceeding as above, we obtain another binomial series:

$$\sum_{k=0}^{\infty} (-1)^k \binom{\alpha}{k} \mathrm{e}^{ksh}$$

that is convergent to the main branch of

$$f(s) = (1 - e^{sh})^{\alpha}$$

provided that Re(s) < 0. This means that the branch cut line of $f(s)$ must be in the left hand half complex plane. We will assume to work in the principal branch and that $f(s)$ is continuous from above. Here we must remark that in $(\cdot)^{\alpha}$ we are again in the principal branch but we are assuming a branch cut line in the right hand complex half plane.

We obtain directly:

$$D_{f}^{\alpha} f(z) = |s|^{\alpha} e^{j\theta\alpha} e^{sz}$$

with $|\theta| < \pi/2$, and

$$D_{b}^{\alpha} f(z) = |s|^{\alpha} e^{j\theta\alpha} e^{sz}$$

valid iff $\theta \in (\pi/2, 3\pi/2)$.

We must be careful in using the above results. In fact, in a first glance, we could be led to use it for computing the derivatives of functions like sin(z), cos(z), sinh(z) and cosh(z). But if we have in mind our reasoning we can conclude immediately that those functions do not have finite derivatives if $z \in \mathbf{C}$. In fact they use simultaneously the exponentials e^{z} and e^{-z} whose derivatives cannot exist simultaneously, as we just saw.

2.7.2 The Constant Function

We start by computing the fractional derivative of the constant function. Let then $f(t) = 1$ for every $t \in \mathbf{R}$ and $\alpha \in \mathbf{R} \backslash \mathbf{Z}$. From (2.32) we have:

$$D_{\theta}^{\alpha} f(t) = \lim_{h \to 0} \frac{\sum_{k=0}^{\infty} (-1)^{k} \binom{\alpha}{k}}{h^{\alpha}} = \begin{cases} 0 & \alpha > 0 \\ \infty & \alpha < 0 \end{cases} \tag{2.57}$$

To prove it, we are going to consider the partial sum of the series

$$\sum_{k=0}^{n} (-1)^{k} \binom{\alpha}{k} = (-1)^{n} \binom{\alpha - 1}{n} = \frac{1}{\Gamma(1-\alpha)} \frac{\Gamma(-\alpha + n + 1)}{\Gamma(n+1)}$$

As $n \to \infty$ the quotient of two gamma functions has a known limiting behaviour[7] [1] that allows us to show that

$$\frac{1}{\Gamma(1-\alpha)} \frac{\Gamma(-\alpha + n + 1)}{\Gamma(n+1)} \to \frac{1}{\Gamma(1-\alpha)} \frac{1}{n^{\alpha}}$$

[7] See Chap. 3.

leading to the limits shown in (2.57). So, the α order fractional derivative of 1 is the null function. If $\alpha < 0$, the limit is infinite. So, there is no fractional "primitive" of a constant. However, this does not happen if α is a negative integer. To exemplify, consider the case $\alpha = -1$. From (2.32), we have:

$$D^{-1}1 = \lim_{L\to\infty} \sum_{n=0}^{L} t/L \tag{2.58}$$

where L is the integer part of t/h. We have

$$D^{-1}1 = t \tag{2.59}$$

It should be stressed that the "primitivation constant" does not appear as expected. This means that when working in the context defined by (2.32) two functions with the same fractional derivative are equal.

The example we just treated allows us to obtain an interesting result:

There are no fractional derivatives of the power function defined in **R** *(or* **C***).*

In fact, suppose that there is a fractional derivative of t^n, $t \in \mathbf{R}$, $n \in \mathbf{N}^+$. We must have:

$$D^{\alpha}t^{+} = n!D^{\alpha}D^{-n}1 = D^{-n}D^{\alpha}1$$

This means that we must be careful when trying to generalise the Taylor series. We conclude also that we cannot compute the fractional derivative of a function by using directly its Taylor expansion. The same result could be obtained directly from (2.32). It is enough to remark that a power function tends to infinite when the argument tends to $-\infty$. The Taylor expansions can be used provided that we consider the causal (right) or anti-causal (left) parts only.

2.7.3 The LT of the Fractional Derivative

The above results can be used to generalize a well known property of the Laplace transform. If we return back to Eq. 2.33 and apply the bilateral Laplace transform

$$F(s) = \int_{-\infty}^{+\infty} f(t)e^{-st}\,ds \tag{2.60}$$

to both sides and use the result in (2.56). We conclude that:

$$\mathrm{LT}\left[D_f^{\alpha}f(t)\right] = s^{\alpha}F(s) \quad \text{for}\,\mathrm{Re}(s) > 0 \tag{2.61}$$

where in s^{α} we assume the principal branch and a cut line in the left half plane. With Eq. 2.34 we obtain:

$$\mathrm{LT}\left[D_b^{\alpha}f(t)\right] = s^{\alpha}F(s) \quad \text{for } \mathrm{Re}(s) < 0 \tag{2.62}$$

where now the branch cut line is in the right half plane. These results have a system interpretation:

There are two systems (differintegrators) with the same expression for the transfer function $H(s) = s^{\alpha}$, but with different regions of convergence: one is causal; the other is anti-causal.

This must be contrasted with the classic integer order case as we referred before. We will not compute the impulse responses here. It will be done later.

2.7.4 The Complex Sinusoid

Now, we are going to see if the above results can be extended to functions with Fourier Transform. We note that the multivalued expression $H(s) = s^{\alpha}$ becomes an analytic function (as soon as we fix a branch cut line) in the whole complex plane excepting the branch cut line. The computation of the derivative of functions with Fourier Transform is dependent on the way used to define $(j\omega)^{\alpha}$. Assume that $H(s)$ is a transfer function with region of convergence defined by $\mathrm{Re}(s) > 0$. This means that we have to choose a branchcut line in the left half complex plane. To obtain the correct definition of $(j\omega)^{\alpha}$ we have to perform the limit as $s \to j\omega$ from the right. We have

$$(j\omega)^{\alpha} = |\omega|^{\alpha} \cdot \begin{cases} e^{j\alpha\pi/2} & \text{if } \omega > 0 \\ e^{-j\alpha\pi/2} & \text{if } \omega < 0 \end{cases} \tag{2.63}$$

If the region of convergence is given by $\mathrm{Re}(s) < 0$, we have to choose a branchcut line in the right half complex plane. If we perform the limit as $s \to j\omega$ from the left we conclude that

$$(j\omega)^{\alpha} = |\omega|^{\alpha} \cdot \begin{cases} e^{j\alpha\pi/2} & \text{if } \omega > 0 \\ e^{j3\alpha\pi/2} & \text{if } \omega < 0 \end{cases} \tag{2.64}$$

We are going to see which consequences impose Eqs. 2.63 and 2.64. They mean that the forward and backward derivatives of a cisoid are given by

$$D_f^{\alpha}e^{j\omega t} = e^{j\omega t}|\omega|^{\alpha} \cdot \begin{cases} e^{j\alpha\pi/2} & \text{if } \omega > 0 \\ e^{-j\alpha\pi/2} & \text{if } \omega < 0 \end{cases} \tag{2.65}$$

and

$$D_b^{\alpha}e^{j\omega t} = e^{j\omega t}|\omega|^{\alpha} \cdot \begin{cases} e^{j\alpha\pi/2} & \text{if } \omega > 0 \\ e^{j3\alpha\pi/2} & \text{if } \omega < 0 \end{cases} \tag{2.66}$$

As the cisoid can be considered as having symmetric behavior in the sense that the segment for $t < 0$ is indistinguishable from the corresponding for $t > 0$, we were expecting that the forward and backward derivatives were equal. This does not happen and the result (2.66) is somehow strange. To see the consequences of this result assume that we want to compute the derivative of a function defined by:

$$f(t) = \frac{1}{2\pi} \int_{-\infty}^{+\infty} F(j\omega) e^{j\omega t} \, d\omega \tag{2.67}$$

we obtain two different inverse transforms, meaning that we have *different transforms with the same domain* and only one of them corresponds to the generalization of a classic property of the Fourier transform: the one obtained with the forward derivative. To reinforce this question let us try to compute the derivative of a real sinusoid: $x(t) = \cos(\omega_0 t)$. From (2.65) we obtain:

$$D_f^\alpha \cos(\omega_0 t) = |\omega_0|^\alpha \cos(\omega_0 t + \alpha\pi/2) \tag{2.68}$$

while, from (2.66) we have:

$$D_b^\alpha \cos(\omega_0 t) = |\omega_0|^\alpha [e^{j\alpha\pi/2} + e^{j3\alpha\pi/2}]/2 \tag{2.69}$$

that is in general a complex function.

These results force us to conclude that:

To compute the fractional derivative of a sinusoid we have to use only the forward derivative.

The Fourier transform of the fractional derivative is computed from the Laplace transform of the forward derivative by computing the limit as s goes to $j\omega$.

We may put the question of what happens with the frequency response of a given fractional linear system. From the conclusions we have just presented, we can say that, having a causal fractional linear system with transfer function equal to $H(s)$, the frequency response must be computed from:

$$H(j\omega) = \lim_{s \to j\omega} H(s) \tag{2.70}$$

This is in agreement with other known results. For example, if the input to the system is white noise, with unit power, the output spectrum is given by:

$$S(\omega) = \lim_{s \to j\omega} H(s)H(-s) \tag{2.71}$$

2.8 Starting from the Transfer Function

To obtain the Grünwald–Letnikov fractional derivative we used a heuristic approach by generalising the integer order derivative defined through the incremental ratio. Here we will show that we can obtain it from the Laplace transform. To do it, it is enough to start from the transfer function $H(s) = s^\alpha$ and express it as a limit as shown in Fig. 2.1

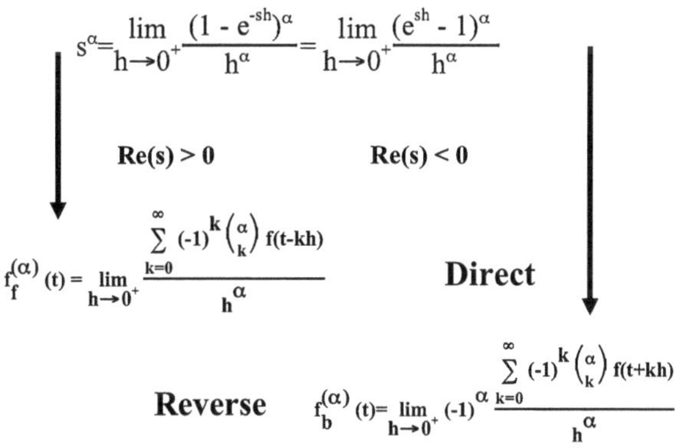

Fig. 2.1 Obtention of the GL derivatives from the transfer function s^{α}

It is not hard to see that s^{α} can be considered as the limit when $h \in R^+$ tends to zero in the left hand sides of the following expressions:

$$\frac{(1 - \mathrm{e}^{-sh})^{\alpha}}{h^{\alpha}} = \frac{1}{h^{\alpha}} \sum_{k=0}^{\infty} (-1)^k \binom{\alpha}{k} \mathrm{e}^{-shk} \quad \mathrm{Re}(s) > 0 \tag{2.72}$$

and

$$\frac{(\mathrm{e}^{sh} - 1)^{\alpha}}{h^{\alpha}} = \frac{(-1)^{\alpha}}{h^{\alpha}} \sum_{k=0}^{\infty} (-1)^k \binom{\alpha}{k} \mathrm{e}^{shk} \quad \mathrm{Re}(s) < 0 \tag{2.73}$$

The right hand sides converge for $\mathrm{Re}(s) > 0$ and $\mathrm{Re}(s) < 0$. This means that the first leads to a causal derivative while the second leads to the anti-causal. These expressions, when inverted back into time lead, respectively, to[8]

$$d_f^{(\alpha)}(t) = \lim_{h\to 0+} \frac{1}{h^{\alpha}} \sum_{k=0}^{\infty} (-1)^k \binom{\alpha}{k} \delta(t - kh) \tag{2.74}$$

and

$$d_b^{(\alpha)}(t) = \lim_{h\to 0+} \frac{(-1)^{\alpha}}{h^{\alpha}} \sum_{k=0}^{\infty} (-1)^k \binom{\alpha}{k} \delta(t + kh) \tag{2.75}$$

Let $f(t)$ be a bounded function and $\alpha > 0$. The convolution of (2.74) and (2.75) with $f(t)$ leads to the Grünwald–Letnikov forward and backward derivatives.

---

[8] We used d instead δ, to avoid confusion, while the equivalence is not proved. We do not care about the convergence of the series [9].

2.9 The Fractional Derivative of Generalized Functions

The results there obtained are valid for analytic functions. Here we will combine the theory there developed with the distribution theory to generalize those results for other functions. We will consider functions defined on R and will treat the forward derivative case, only by simplicity, omitting the subscript. However, we will deal with functions that not only they are not analytic, but they can be discontinuous. This leads us to the Distribution (Generalized Function) Theory. We will use the results of the Axiomatic Theory due to its simplicity. Combining both results we will show that the main interesting properties of the above derivative we obtained for analytic functions remain valid in a distributional context.

The above properties are valid provided that all the involved derivatives exist. This may not happen in a lot of situations; for example, the derivative of the power function causal or not. As these functions are very important we will consider them with detail. Meanwhile let us see how we can enlarge the validity of the above formulae. Let us consider formula (2.33), because the others can be treated immediately from it.

Consider a function $f(t)$ such that there exists $D^{\alpha}f(t)$ but it is not continuous. In principle, we cannot assure that we can apply (2.33) to obtain $D^{\alpha+\beta}f(t)$. To solve the problem, we will use a suitable definition of distribution. Due to its simplicity, we adopt here the definition underlying the Axiomatic Theory of Distributions [9–12]. It states that: *a Distribution is an integer order derivative of a continuous function.*

Consider then that $f(t) = D^n g(t)$, where n is a positive integer and $g(t)$ is continuous and with continuous fractional derivative of order $\alpha + \beta$. In this case, we can write:

$$D^{\alpha+\beta}f(t) = D^{\alpha+\beta}D^n g(t) = D^n D^{\alpha+\beta}g(t)$$

So, we obtained the desired derivative by integer order derivative computation of the fractional derivative. The other properties are consequence of this one. Some examples will clarify the situation.

2.9.1 The Causal Power Function

The results obtained in the above close section allow us to obtain the derivative of any order of the function $p(t) = t^{\beta}u(t)$, with $\beta > 0$. It is a continuous function, thus indefinitely (integer order) derivable. To compute the fractional derivative of $p(t)$, the easy way is to use the Laplace transform (LT). As well known, the LT of $p(t)$ is $P(s) = \frac{\Gamma(\beta+1)}{s^{\beta+1}}$, for $\text{Re}(s) > 0$. The transform of the fractional derivative of order α is given by: $s^{\alpha}\frac{\Gamma(\beta+1)}{s^{\beta+1}}$. So,

$$D_f^{\alpha} t^{\beta} u(t) = \frac{\Gamma(\beta+1)}{\Gamma(\beta-\alpha+1)} t^{\beta-\alpha} u(t) \tag{2.76}$$

that generalizes the integer order formula for α, $\beta \in \mathbf{R}^+$ and $\beta > \alpha$ [1, 13, 14]. In fact, if $\alpha = N$, we obtain

$$D_f^{N} t^{\beta} u(t) = (\beta)_N t^{\beta-N} u(t) \tag{2.77}$$

that is the result we obtain by successive order one derivations. To obtain it, we use the rule of the derivative of the product. The derivative of $u(t)$ is $\delta(t)$ that appears multiplied by a power that is zero at $t = 0$ [9]

$$Dt^{\beta} u(t) = \beta t^{\beta-1} u(t) + t^{\beta} \delta(t) = \beta t^{\beta-1} u(t)$$

Equation 2.76 can be considered valid for $\beta - \alpha = -1$ provided that we write

$$D_f^{\alpha} t^{\alpha-1} u(t) = \delta(t) \tag{2.78}$$

If $\beta = N \in \mathbf{Z}^+$, we have:

$$D_f^{\alpha}[t^N u(t)] = \frac{N!}{\Gamma(N+1-\alpha)} [t^{N-\alpha} u(t)] \tag{2.79}$$

if $N > \alpha$. This result shows that the derivative of a given causal function can be computed from the causal McLaurin series by computing the derivatives of the series terms. This means that we can obtain the derivative of the causal exponential $e^{at}u(t)$ by computing the derivative of each term of the McLaurin series. However, the resulting series is not easily related to the exponential. This will be done later.

Let as return to our initial objective: to enlarge the validity of (2.76). With the assumed values for α and β, (2.76) represents a continuous function. So, we can compute the Nth order derivative. Again products of powers and the $\delta(t)$ will appear, but now the powers assume an infinite value at $t = 0$. We remove this term to obtain what is normally called *finite part of the distribution*. If $\gamma < 0$, we have

$$Dt^{\gamma} u(t) = \gamma t^{\gamma-1} u(t) + t^{\gamma} \delta(t)$$

The second term on the right is removed to give us the finite part [9, 12]

$$F_p[Dt^{\gamma} u(t)] = \gamma t^{\gamma-1} u(t)$$

In the following, we will assume that we will be working with the finite part and will omit the symbol F_p. With these considerations we conclude that (2.76) remains valid provided that $\alpha \in R$ and $\beta \in R - Z^-$. In particular, we have:

$$D_f^{\alpha} u(t) = \frac{1}{\Gamma(1-\alpha)} t^{-\alpha} u(t) \tag{2.80}$$

To confirm the validity of the procedure, we are going to obtain this result directly from (2.33). For $t < 0$, $D^{\alpha}u(t)$ is zero; for $t > 0$, we have

$$D_f^{\alpha}u(t) = \lim_{h\to 0+} \frac{\sum_{k=0}^{L}(-1)^k\binom{\alpha}{k}}{h^{\alpha}} \tag{2.81}$$

where L is the integer part of t/h: $L = \lfloor t/h \rfloor$ and α is a positive non-integer real (otherwise it leads to $\delta(t)$ and its derivatives). We are going to make some manipulations to obtain the required result. If h is very small, $L \approx t/h$ and we have:

$$D_f^{\alpha}u(t) = \lim_{h\to 0+} \frac{\sum_{k=0}^{L}\frac{(-\alpha)_k}{k!}}{h^{\alpha}} = \lim_{L\to\infty} t^{-\alpha}L^{\alpha}\sum_{k=0}^{L}\frac{(-\alpha)_k}{k!} \tag{2.82}$$

To go further we use one interesting property of the Gauss hypergeometric function ([15]; Wolfram.com):

$${}_2F_1(-n,b,-m,z) = \sum_0^n \frac{(-n)_k(b)_k z^k}{(-m)_k k!}$$

where n, $m \in Z^+$, with $m \geq n$. Putting $z = 1$, $b = -\alpha$, and $m = n$, we can write:

$$D_f^{\alpha}u(t) = \lim_{L\to\infty} t^{-\alpha}L^{\alpha}_2F_1(-L,-\alpha,-L,1) \tag{2.83}$$

But

$${}_2F_1(a,b,c,1) = \frac{\Gamma(c)\Gamma(c-a-b)}{\Gamma(c-a)\Gamma(c-b)}$$

if $\mathrm{Re}(c - a - b) > 0$. The application of this formula leads to an undetermination

$${}_2F_1(-L,-\alpha-L,1) = \frac{\Gamma(-L)\Gamma(\alpha)}{\Gamma(\alpha-L)\Gamma(0)} \quad \alpha > 0$$

However attending to the residues of the gamma function at the poles, we can write:

$$\frac{\Gamma(-L)}{\Gamma(0)} = \frac{(-1)^L}{L!}$$

and

$${}_2F_1(-L,-\alpha-L,1) = \frac{(-1)^L}{L!}\frac{\Gamma(\alpha)}{\Gamma(\alpha-L)} = \frac{(1-\alpha)_L}{L!} = \frac{(-\alpha)_{L+1}}{-\alpha L!} \tag{2.84}$$

With this result we can write (2.83) as

$$D_f^{\alpha}u(t) = t^{-\alpha}\lim_{L\to\infty} L^{\alpha}\frac{(-\alpha)_{L+1}}{-\alpha L!} \tag{2.85}$$

In the right hand side we recognize the gamma function [16] leading to the expected result. If α is a negative integer, $-N$, we know that

$$D_f^{-N}u(t) = \frac{t^N}{N!}u(t) \tag{2.86}$$

and attending to (2.86) we conclude that (2.76) is valid for $\alpha \in R - Z^+$. Equation 2.80 allows us to obtain the interesting result

$$D_f^{\alpha}\delta(t) = \frac{t^{-\alpha-1}}{\Gamma(-\alpha)}u(t) \tag{2.87}$$

valid for positive non-integer orders. In terms of linear system theory, (2.87) tells us that the fractional forward differintegrator (a current terminology) is a linear system with impulse response equal to the right hand side. As the output is given by the convolution of the input and the impulse response, we obtain [6, 17]

$$D_f^{\alpha}f(t) = \frac{1}{\Gamma(-\alpha)}\int_{-\infty}^{t} f(\tau)(t-\tau)^{-\alpha-1}\,d\tau \tag{2.88}$$

that we will call the forward Liouville derivative.

Using (2.33) and (2.87), we can obtain a very curious result

$$\frac{t^{-\alpha-1}}{\Gamma(-\alpha)}u(t) = \lim_{h\to 0+}\frac{\sum_{k=0}^{\infty}(-1)^k\binom{\alpha}{k}\delta(t-kh)}{h^{\alpha}} \tag{2.89}$$

meaning that the power function is obtained by joining infinite impulses with infinitesimal amplitudes modulated by the binomial coefficients.

A similar procedure would lead us to obtain the impulse response of the backward differintegrator that is given by [6, 17]

$$D_b^{\alpha}\delta(t) = -\frac{t^{-\alpha-1}u(-t)}{\Gamma(-\alpha)} \tag{2.90}$$

that allows us to obtain the backward Liouville derivative

$$D_b^{\alpha}f(t) = -\frac{1}{\Gamma(-\alpha)}\int_{t}^{\infty} f(\tau)\cdot(t-\tau)^{-\alpha-1}\,d\tau \tag{2.91}$$

With a change of variable it leads to

$$D_b^{\alpha}f(t) = -\frac{(-1)^{-\alpha}}{\Gamma(-\alpha)}\int_{0}^{\infty} f(t+\tau)\cdot\tau^{-\alpha-1}\,d\tau \tag{2.92}$$

that shows the anti-causal characteristic on depending on the future values of the function.

These integral formulations were introduced both exactly with this format by Liouville. Unhappily in the common literature the factor $(-1)^{-\alpha}$ in (2.92) has been removed and is called Weyl derivative [18]. Although the above results were obtained for functions with Laplace transform their validity can be extended to other functions.

Integrals (2.88) and (2.91) are not very useful since they are "hyper singular" integrals [1, 13].

2.9.2 The Causal Exponential

The derivative of the causal exponential can be obtained from the McLaurin series, as we said above. We have:

$$D^{\alpha}[e^{at}u(t)] = \sum_{0}^{\infty} \frac{(at)^{k-\alpha}}{\Gamma(k-\alpha)} u(t) \tag{2.93}$$

However, this function is not very interesting, since it does not represents the fractional generalization of the causal exponential. To obtain it, put $\beta = n\alpha$ in (2.76) and rewrite it in the format:

$$D^{\alpha} \frac{t^{n\alpha}u(t)}{\Gamma(n\alpha+1)} = \frac{1}{\Gamma((n-1)\alpha+1)} t^{(n-1)\alpha} u(t) \tag{2.94}$$

We are led to the Mittag–Leffler function:

$$g(t) = \sum_{n=0}^{\infty} \frac{t^{n\alpha}}{\Gamma(n\alpha+1)} u(t) \tag{2.95}$$

It is not hard to show, using (2.94), that

$$D^{\alpha} g(t) = \sum_{n=0}^{\infty} \frac{t^{n\alpha}}{\Gamma(n\alpha+1)} \cdot u(t) + \frac{t^{-\alpha}}{\Gamma(-\alpha+1)} u(t) \tag{2.96}$$

or with (2.80)

$$D^{\alpha}[g(t) - u(t)] = g(t)$$

This means that $g(t)$ is the solution of the fractional differential equation

$$D^{\alpha} g(t) - g(t) = 0 \tag{2.97}$$

under the initial condition $g(0) = 1$.

2.9.3 The Causal Logarithm

We going to study the causal logarithm $\lambda(t) = log(t) \cdot u(t)$ We could use Eq. 2.32, but the computations are somehow involved. Instead, we start from relation (2.76) and compute the derivative of both sides relative to β to obtain:

$$D^{\alpha}[t^{\beta}\log(t)u(t)] = \frac{\Gamma(\beta+1)}{\Gamma(\beta-\alpha+1)}t^{\beta-\alpha}u(t)[\log(t)+\psi(\beta+1)-\psi(\beta-\alpha+1)] \tag{2.98}$$

where we represented by ψ the logarithmic derivative of the gamma function:

$$\psi(t) = D[\log\Gamma(t)] = \Gamma'(t)/\Gamma(t)$$

Putting $\beta = 0$ in (2.98) we obtain the derivative of the causal logarithm

$$D^{\alpha}[\log(t)u(t)] = \frac{1}{\Gamma(-\alpha+1)}t^{-\alpha}u(t)[\log(t)-\gamma-\psi(-\alpha+1)] \tag{2.99}$$

where $\gamma = -\psi(1) = -\Gamma'(1)$ is the Euler–Mascheroni constant.

Another interesting result can be obtained by integer order derivation of both sides in (2.99). As

$$D^{N}[\log(t)u(t)] = (-1)^{N-1}(N-1)!t^{-N}u(t), \quad N = 1,2,\ldots \tag{2.100}$$

we have:

$$\begin{aligned} D^{\alpha}t^{-N}u(t) &= \frac{D^{N}[\log(t)u(t)]}{(-1)^{N-1}(N-1)!} \\ &= \frac{1}{(-1)^{N-1}(N-1)!\Gamma(-\alpha-N+1)}t^{-\alpha-N}u(t)[\log(t)-\gamma-\psi(-\alpha-N+1)] \end{aligned} \tag{2.101}$$

From the properties of the gamma functions, we have:

$$\Gamma(-\alpha-N+1) = (-1)^{N}\frac{\Gamma(1-\alpha)}{(\alpha)_N}$$

and using repeatedly the recurrence property of the digamma we obtain:

$$\psi(-\alpha-N+1) = \psi(-\alpha+1) - \sum_{n=1}^{N}\frac{1}{1-\alpha-n}$$

leading to

$$D^{\alpha}t^{-N}u(t) = -\frac{(\alpha)_N}{(N-1)!\Gamma(-\alpha+1)}t^{-\alpha-N}u(t)\left[\log(t) - \gamma - \psi(-\alpha+1) - \sum_{n=1}^{N}\frac{1}{1-\alpha-n}\right] \tag{2.102}$$

If $\alpha = M$ is a positive integer, the gamma function $\Gamma(-\alpha + 1)$ is infinite and we have:

$$D^{M}t^{-N}u(t) = \lim_{\alpha\to M}\frac{\psi(-\alpha+1)}{\Gamma(-\alpha+1)}\frac{(\alpha)N}{(N-1)!}t^{-\alpha-N}u(t) \tag{2.103}$$

But

$$\Gamma(x) \approx (-1)^n\frac{1}{n!(x+n)}$$

near the pole at $x = -n$. Then

$$\lim_{\alpha\to M}\frac{\psi(-\alpha+1)}{\Gamma(-\alpha+1)} = (-1)^M(M-1)!$$

and, finally

$$D^{M}t^{-N}u(t) = (-1)^M(M)_N t^{-M-N}u(t) \tag{2.104}$$

that is the classic result. It is not very difficult to obtain this result from (2.76) provided that we take care about the indetermination. To conclude:

$$D_f^{\alpha}t^{\beta}u(t) = \begin{cases} \frac{\Gamma(\beta+1)}{\Gamma(\beta-\alpha+1)}t^{\beta-\alpha}u(t) & \beta \in R - Z^- \\ -\frac{(\alpha)_N}{(N-1)!\Gamma(-\alpha+1)}t^{-\alpha-N}u(t)\left[\log(t) - \gamma - \psi(-\alpha+1) - \sum_{n=1}^{N}\frac{1}{1-\alpha-n}\right] & -N = \beta \in Z^- \end{cases} \tag{2.105}$$

We must remark that these results are more general than those obtained by integral derivative formulations like Riemann–Liouville or Caputo derivatives [13, 14].

2.9.4 Consequences in the Laplace Transform Domain

Consider that we are working in the context of the Laplace transform. With the two-sided LT we use there are no problems in defining the LT of $\delta(t)$:

$$\mathrm{LT}[\delta(t)] = 1$$

Attending to (2.61) and (2.87) we conclude that

$$\mathrm{LT}\left[\frac{t^{-\alpha-1}}{\Gamma(-\alpha)}u(t)\right] = s^{\alpha}, \quad \mathrm{Re}(s) > 0$$

that generalizes a well known result for any $\alpha \in R - Z^{-}$. Essentially, we prolonged the sequence:

$$\ldots s^{-n} \ldots, s^{-2}, s^{-1}, 1, s^{1}, s^{2}, \ldots, s^{n} \ldots$$

in order to include other kinds of exponents: rational, real, or even complex numbers. In agreement with what we said before, there are two forms of obtaining the extension, depending on the choice done for region of convergence for the LT: the left and right half-planes. This has some implications in the study of the fractional linear systems as we will see later.

From (2.76) we obtain easily the LT of the Mittag–Leffler function

$$G(s) = \sum_{n=0}^{\infty} \frac{1}{s^{n\alpha+1}} = \frac{s^{\alpha-1}}{s^{\alpha}-1}; \quad \mathrm{Re}(s) > 1 \tag{2.106}$$

Relation (2.106) expresses a special case of the more general result known as Hardy's theorem [16] that states:

Let the series

$$F(s) = \sum_{0}^{+\infty} a_n \Gamma(\alpha + n + 1) \cdot s^{-\alpha-n-1} \tag{2.107}$$

be convergent for some $\mathrm{Re}(s) > s_0 > 0$ *and* $\alpha > -1$. *The series*

$$f(t) = \sum_{0}^{+\infty} a_n t^{\alpha+n} \tag{2.108}$$

converges for all $t > 0$ *and* $F(s) = \mathrm{LT}[f(t)]$.

In agreement with the results in Sect. 2.9.1, the validity of the Hardy's theorem can be extended to values of $\alpha < -1$, provided it is not integer.

2.10 Riemann–Liouville and Caputo Derivatives

The Riemann–Liouville and Caputo derivatives are multistep derivatives that use several integer order derivatives and a fractional integration [1, 13, 14, 19]. Although they create some difficulties as we will see later, we are going to describe them, since they are widely used with questionable results. To present them, we use (2.87) and (2.90) to obtain the following distributions:

$$\delta_{\pm}^{(-\nu)}(t) = \pm\frac{t^{\nu-1}}{\Gamma(\nu)}u(\pm t), \quad 0<\nu<1$$

and

$$\delta_{\pm}^{(n)}(t) = \begin{cases} \pm\frac{t^{-n-1}}{(-n-1)!}u(\pm t) & \text{for } n<0 \\ \delta^{(n)}(t) & \text{for } n \geq 0 \end{cases} \tag{2.109}$$

where $n \in \mathbf{Z}$. With them we define two differintegrations usually classified as left and right sided, respectively:

$$f_l^{(\alpha)}(t) = [f(t)u(t-a)] * \delta_{+}^{(n)}(t) * \delta_{+}^{(-\nu)}(t) \tag{2.110}$$

$$f_r^{(\alpha)}(t) = [f(t)u(b-t)] * \delta_{+}^{(n)}(-t) * \delta_{+}^{(-\nu)}(-t) \tag{2.111}$$

with $a < b \in \mathbf{R}$. The orders are given by $\alpha = n - \nu$, n being the least integer greater than α and $0 < \nu < 1$. In particular, if α is integer then $\nu = 0$.[9] From different orders of commutability and associability in the double convolution we can obtain distinct formulations. For example, from (2.111) we obtain the left Riemann–Liouville and Caputo derivatives:

$$f_{RL+}^{(\beta)}(t) = \delta_{+}^{(n)}(t) * \left\{[f(t)u(t-a)] * \delta_{+}^{(-\nu)}(t)\right\} \tag{2.112}$$

$$f_{C+}^{(\beta)}(t) = \left\{[f(t)u(t-a)] * \delta_{+}^{(n)}(t)\right\} * \delta_{+}^{(-\nu)}(t) \tag{2.113}$$

For the right side derivatives the procedure is similar.

We are going to study more carefully the characteristics of these derivatives. Consider (2.113). Let

$$\varphi^{(-\nu)}(t) = \left\{[f(t)u(t-a)] * \delta_{+}^{(-\nu)}(t)\right\}.$$

We have:

$$\varphi^{(-\nu)}(t) = \begin{cases} \frac{1}{\Gamma(\nu)}\int_a^t f(\tau)\cdot(t-\tau)^{\nu-1}\,\mathrm{d}\tau & \text{if } t > a \\ 0 & \text{if } t < a \end{cases}$$

So, in general when doing the second convolution in (2.113) we are computing the integer order derivative of a function with a jump. The "jump formula" [9, 10][10] leads to

$$f_{RL+}^{(\beta)}(t) = \frac{1}{\Gamma(-\alpha)}\int_a^t f(t)\cdot(t-\tau)^{-\alpha-1}\,\mathrm{d}\tau - \sum_{i=0}^{n-1} f^{(\alpha-1-i)}(a)\delta^{(i)}(t) \tag{2.114}$$

---

[9] All the above formulae remain valid in the case of integer integration, provided that we put $\delta^{(0)}(t) = \delta(t)$.

[10] It will be studied later.

The appearance of the "initial conditions" $f^{(\alpha-1-i)}(a+)$ provoked some confusions because they were used as initial conditions of linear systems. This is not correct in general. They represent what we need to join to the Riemann–Liouville derivative to obtain the Liouville derivative (2.88). We will return to this subject when we study the initial condition problem in linear fractional systems. Now let us do a similar analysis to the Caputo derivative. The expression

$$\left\{[f(t)u(t-a)] * \delta_+^{(n)}(t)\right\}$$

states the integer order derivative of the function $f(t)\cdot u(t-a)$ Again the jump formula gives

$$y^{(n)}(t)\cdot u(t-a) = [y(t)\cdot u(t-a)]^{(n)} - \sum_{i=0}^{n-1} y^{(n-1-i)}(a)\delta^{(i)}(t) \tag{2.115}$$

that leads to:

$$f_{C+}^{(\beta)}(t) = \frac{1}{\Gamma(-\alpha)}\int_a^t f(\tau)\cdot(t-\tau)^{-\alpha-1}\,\mathrm{d}\tau - \sum_{i=0}^{n-1} f^{(n-1-i)}(a)\delta^{(i-v)}(t) \tag{2.116}$$

In this case, we can extract conclusions similar to those we did in the Riemann–Liouville case. Relation (2.116) explains why sometimes the first n terms of the Taylor series of $f(t)$ are subtracted to it before doing a fractional derivative computation. It is like a regularization.

2.11 The Fractional Derivative of Periodic Signals

2.11.1 Periodic Functions

Let us consider a function defined by Fourier series

$$f(t) = \sum_{-\infty}^{+\infty} F_n e^{j2\pi/Tnt} \tag{2.117}$$

Only by simplicity we will assume for now that the series is uniformly convergent almost everywhere over the real straight line. As it is well known, it defines a almost everywhere continuous periodic function, $f(t)$

$$f(t) = f(t+T), \quad t\in R \tag{2.118}$$

where T is the period and we can write

$$f(t) = \sum_{-\infty}^{+\infty} f_\beta(t - nT) \tag{2.119}$$

where $f_b(t)$ is an almost everywhere continuous function with bounded support. It is a simple task to see that it can be expressed by the convolution

$$f(t) = f_b(t) * \sum_{-\infty}^{+\infty} \delta(t - nT) \tag{2.120}$$

We obtained three different was of representing a periodic function.

The results presented in Sect. 2.7.4 show that we can obtain the derivative of a periodic function from its Fourier series by computing the derivative term by term. We obtain:

$$D_f^\alpha f(t) = \left(\frac{2\pi j}{T}\right)^\alpha \sum_{-\infty}^{+\infty} n^\alpha F_n e^{j2\pi/Tnt} \tag{2.121}$$

It would be interesting to study the convergence of this series, but it is not important here. We can use the procedure followed in study of the comb signal [20] that was an adaptation of the theory presented in [9].

We concluded that, if a periodic function is defined by it Fourier series it can be fractionally derivated term by term and the derivative is also periodic with the same period.

Now we go back to (2.119). As the derivative is a linear operator, we can apply it to each term of the series. Besides $f_b(t)$ is a function with bounded support which implies that the summation in (2.32) is done over a finite number of parcels. So, it is convergent for each h and has a fractional derivative. Another way of concluding this is by using the Laplace transform and (2.49). This leads to

$$D_f^\alpha f(t) = \sum_{-\infty}^{+\infty} D_f^\alpha f_\beta(t - nT) \tag{2.122}$$

However, the fractional derivative of a bounded support function does not have bounded support. This means that the series of the derivatives has what is called in signal processing a aliasing which may prevent its convergence. Anyway, the fractional derivative of a periodic function defined by (2.119) exists (at least formally) and is a periodic function with the same period.

Finally, let us consider the representation (2.120). The fractional derivative will be given by the convolution of $f_b(t)$ and the fractional derivative of the comb function [20]. Attending to (2.87) we conclude that the derivative of the comb is

$$D_f^\alpha \sum_{-\infty}^{+\infty} \delta(t - nT) = \frac{1}{\Gamma(-\alpha)} \sum_{-\infty}^{+\infty} (t - nT)^{-\alpha-1} u(t - nT) \tag{2.123}$$

that convolved with $f_b(t)$ leads again to (2.122).

We are going now to study the causal periodic case and show that the fractional-order derivatives of a causal periodic function with a specific period cannot be a periodic function with the same period. Consider relation (2.120) again and assume that the support of $f_b(t)$ is the interval (0, T). It is immediate to verify that the causal periodic function is given by

$$g(t) = \sum_{0}^{+\infty} f_\beta(t - nT) \tag{2.124}$$

or equivalently:

$$g(t) = f(t) \cdot u(t) \tag{2.125}$$

Assume that $0 < \alpha < 1$. We are going to use the Caputo derivative only with illustration objective. The same result could be obtained with the GL or Liouville derivatives. In a first step, we obtain:

$$g'(t) = f'(t) \cdot u(t) + f(0) \cdot \delta(t) \tag{2.126}$$

and

$$D^\alpha g(t) = \frac{1}{\Gamma(1-\alpha)} \int_0^t f'(\tau) \cdot (t-\tau)^{-\alpha} \, \mathrm{d}\tau + f(0) \cdot t^{-\alpha} u(t) \tag{2.127}$$

We conclude that $D^\alpha g(t)$ cannot be periodic, because the second term destroys any hypothesis of periodicity. However, we can always remove a constant function without affecting the periodicity. So, without loosing generality, let us assume that $f(0) = 0$, for example that $f(t) = \sin(\omega_0 t)$. We can write:

$$\begin{aligned} D^\alpha g(t) &= \frac{\omega_0}{\Gamma(1-\alpha)} \int_0^t \cos(\omega_0 \tau) \cdot (t-\tau)^{-\alpha} \, \mathrm{d}\tau = \frac{\omega_0}{\Gamma(1-\alpha)} \int_0^t \cos[\omega_0 (t-\tau) \cdot \tau^{-\alpha} \, \mathrm{d}\tau] \\ &= \frac{2\omega_0}{\Gamma(1-\alpha)} \mathrm{Re} \left\{ \mathrm{e}^{j\omega_0 t} \int_0^t \mathrm{e}^{-j\omega_0 \tau} \cdot \tau^{-\alpha} \, \mathrm{d}\tau \right\} \end{aligned} \tag{2.128}$$

Take the above integral and write

$$\begin{aligned} \frac{1}{\Gamma(1-\alpha)} \int_0^t \mathrm{e}^{-j\omega_0 \tau} \cdot \tau^{-\alpha} \, \mathrm{d}\tau = & \frac{1}{\Gamma(1-\alpha)} \int_0^\infty \mathrm{e}^{-j\omega_0 \tau} \cdot \tau^{-\alpha} \, \mathrm{d}\tau \\ & - \frac{1}{\Gamma(1-\alpha)} \int_t^\infty \mathrm{e}^{-j\omega_0 \tau} \cdot \tau^{-\alpha} \, \mathrm{d}\tau \end{aligned}$$

The first term is the Fourier transform of $t^{-a}u(t)$. Using the results of Sect. 2.7.4 it assumes the value $(j\omega_0)^{\alpha-1}$. We obtain from (2.61)

$$D^\alpha g(t) = |\omega_0|^\alpha \sin(\omega_0 t + \alpha\pi/2)u(t) - \frac{2\omega_0}{\Gamma(1-\alpha)}\mathrm{Re}\left\{ e^{j\omega_0 t} \int_t^\infty e^{-j\omega 0\tau} \cdot \tau^{-\alpha}\, d\tau \right\} \tag{2.129}$$

We conclude that $D^\alpha g(t)$ in not causal periodic, because there is a transient term. We could also obtain this result by applying the generalized Leibniz rule to $g(t)$.

2.12 Fractional Derivative Interpretations

There have been several attempts to give an interpretation (normally, geometric) to the fractional derivative. The most interesting were proposed by Podlubny [18] and by Machado [9, 10]. The first proposed geometric interpretations for the integral formulations of the fractional derivative. The second gives a probabilistic interpretation for the GL derivative. These attempts although interesting were not widely accepted and did not give rise to new problem solving methodology.

2.13 Conclusions

We approached the fractional derivative definition through a generalized Grünwald–Letnikov formulation. We presented the most interesting properties it enjoys namely the causality and the group properties. We deduced integral formulations and obtained the called Riemann–Liouville and Caputo derivatives. We considered the derivative of periodic signals.

References

1. Samko SG, Kilbas AA, Marichev OI (1987) Fractional integrals and derivatives—theory and applications. Gordon and Breach Science Publishers, New York
2. Hoskins RF (1999) Delta functions. Horwood Series in Mathematics & Applications, Chichester, England
3. Dugowson S (1994) Les différentielles métaphysiques. PhD thesis. Université Paris Nord
4. Diaz JB, Osler TJ (1974) Differences of fractional order. Math Comput 28(125):185–202
5. Ortigueira MD (2004) From differences to differintegrations. Fractional Calculus Appl Anal 7(4):459–471
6. Ortigueira MD (2006) A coherent approach to non integer order derivatives. Signal processing special section: fractional calculus applications in signals and systems, vol 86(10). pp 2505–2515

7. Oldham KB, Spanier J (1974) The fractional calculus: theory and application of differentiation and integration to arbitrary order. Academic Press, New York
8. Krempl PW Solution of linear time invariant differential equations with ‘proper’ primitives. IEEE 32nd Annual Conference on Industrial Electronics, IECON 2006, 6-10 Nov, Paris, France
9. Ferreira JC (1997) Introduction to the theory of distributions. Pitman monographs and surveys in pure and applied mathematics. Longman, Harlow
10. Hoskins RF, Pinto JS (1994) Distributions, ultradistributions, and other generalised functions. Ellis Horwood Limited, Chichester
11. Silva JS (1989) The axiomatic theory of the distributions. Complete Works, INIC, Lisbon
12. Zemanian AH (1987) Distribution theory and transform analysis. Dover Publications, New York
13. Kilbas AA, Srivastava HM, Trujillo JJ (2006) Theory and applications of fractional differential equations. Elsevier, Amsterdam
14. Podlubny I (1999) Fractional differential equations: an introduction to fractional derivatives, fractional differential equations, to methods of their solution and some of their applications. Academic Press, San Diego
15. Chaudhry MA, Zubair SM (2002) On a class of incomplete gamma functions with applications. Chapman & Hall, CRC, London, Boca Raton
16. Henrici P (1991) Applied and computational complex analysis, vol 2. Wiley, New York, pp 389–391
17. Ortigueira MD (2000) Introduction to fractional signal processing. Part 1: continuous-time systems. IEE Proc Vis Image Signal Process 1:62–70
18. Miller KS, Ross B (1993) An introduction to the fractional calculus and fractional differential equations. Wiley, New York
19. Magin RL (2006) Fractional calculus in bioengineering. Begell House, Connecticut
20. Ortigueira MD (2001) The comb signal and its Fourier transform. Signal Process 81(3):581–592
21. Podlubny I (2002) Geometric and physical interpretation of fractional integration and fractional differentiation. J Fract Calc Appl Anal 5(4):367–386
22. Machado JAT (2003) A probabilistic interpretation of the fractional-order differentiation. Fractional Calculus Appl Anal 6(1):73–80
23. Machado JAT (2009) Fractional derivatives: probability interpretation and frequency response of rational approximations. Commun Nonlinear Sci Numer Simulat 14:3492–3497

Chapter 3
Integral Representations

3.1 Introduction

In the previous chapter we addressed the problem of fractional derivative definition and proposed the use the Grünwald–Letnikov and in particular the forward and backward derivatives. These choices were motivated by five main reasons they:

- do not need superfluous derivative computations,
- do not insert unwanted initial conditions,
- are more flexible,
- allow sequential computations,
- are more general in the sense of allowing to be applied to a large class of functions.

We presented also the Liouville derivatives that we deduced from the convolutional property of the Laplace transform.

However there are other integral representations mainly when working in a complex setup. It is well known that the formulation in the complex plane is represented by the generalised Cauchy derivative. So, we need a coherent mathematical reasoning for a connection between the GL formulation and the generalised Cauchy. We are going to present it.

In facing this problem, we assume here as starting point the definitions of direct and reverse fractional differences and present their integral representations. From these representations and using the asymptotic properties of the Gamma function, we will obtain the generalised Cauchy integral as a unified formulation for any order derivative in the complex plane. As we will see

---

In a first reading, people more interested in applications can jump this chapter.

M. D. Ortigueira, *Fractional Calculus for Scientists and Engineers*, Lecture Notes in Electrical Engineering, 84, DOI: 10.1007/978-94-007-0747-4_3,

The generalised Cauchy derivative of analytic functions is equal to the Grünwald–Letnikov fractional derivative.

When trying to compute the Cauchy integral using the Hankel contour we conclude that:

- The integral have two terms: one corresponds to a derivative and the other to a primitive.
- The exact computation leads to a regularized integral, generalising the well known concept of pseudo-function, but without rejecting any infinite part.
- The definition implies causality.

The forward and backward derivatives emerge again as very special cases. We will study them for the case of functions with Laplace Transforms. This leads us to obtain once again the causal and anti-causal fractional linear differintegrators both with Transfer Function equal to s^{α}, $\alpha \in \mathbf{R}$ in agreement with the mathematical development presented in Chap. 2.

3.2 Integral Representations for the Differences

In Chap. 2, we presented the general descriptions of fractional differences. These were based on a study by Diaz and Osler [1]. They proposed an integral formulation for the differences and conjectured about the possibility of using it for defining a fractional derivative. This problem was also discussed in a round table held at the International Conference on "Transform Methods & Special Functions", Varna'96 as stated by Kiryakova [2]. The validity of such conjecture was proved [3, 4] and used to obtain the Cauchy integrals from the differences and simultaneously generalise it to the fractional case. Those integral formulations for the fractional differences will be presented in the following. We will start by the integer order case.

3.2.1 Positive Integer Order

We return back to Sect. 2.2.1 and recover the formulae for the differences we presented there. Consider first the positive integer order case (2.11) and (2.12). Assume that $f(z)$ is analytic inside and on a closed integration path that includes the points $t = z - kh$ in the direct case and $t = z + kh$ in the corresponding reverse case, with $k = 0, 1, \ldots, N$ {see Fig. 3.1} and $Re(h) > 0$.

The results stated in (2.11) and (2.12) can be interpreted in terms of the residue theorem.[1] In fact they can be considered as $\frac{1}{2\pi j}\cdot \sum R_i$ where R_i $i = 1, 2, \ldots$ are the

[1] It is important to remark that the poles are simple and that this case can be deduced without the use of the derivative notion.

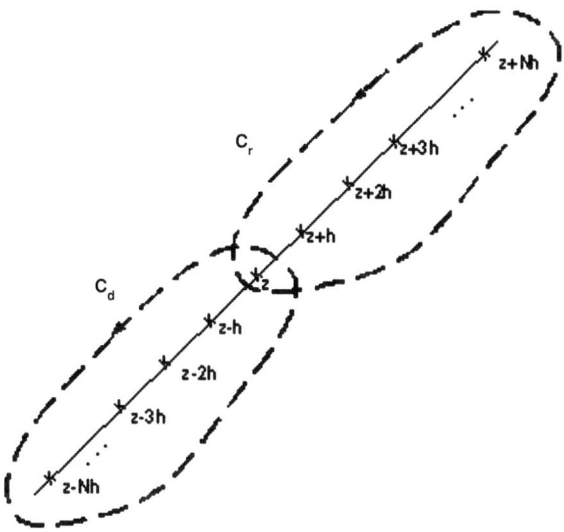

Fig. 3.1 Integration paths and poles for the integral representation of integer order differences

residues in the computation of the integral of a function with poles at $t = z - kh$ and $t = z + kh$, $k = 0, 1, 2, \ldots$ As it can be seen by direct verification, we have:

$$\sum_{k=0}^{N}(-1)^k\binom{N}{k}f(z-kh) = \frac{N!}{2\pi jh}\int_{C_d}\frac{f(w)}{\prod_{k=0}^{N}\left(\frac{w-z}{h}+k\right)}\mathrm{d}w \tag{3.1}$$

and

$$\sum_{k=0}^{N}(-1)^k\binom{N}{k}f(z+kh) = \frac{N!}{-2\pi jh}\int_{C_r}\frac{f(w)}{\prod_{k=0}^{N}\left(\frac{z-w}{h}+k\right)}\mathrm{d}w \tag{3.2}$$

We must remark that the binomial coefficients appear naturally when computing the residues. These formulations are more general than those proposed by Diaz and Osler, because they considered only the $h = 1$ case.

The product in the denominator in the above formulae is called shifted factorial and is usually represented by the Pochhammer symbol. With it we can express the differences in the following integral formulations:

$$\Delta_d^N f(z) = \frac{N!}{2\pi jh}\int_{C_d}\frac{f(w)}{\left(\frac{w-z}{h}\right)_{N+1}}\mathrm{d}w \tag{3.3}$$

and

$$\Delta_r^N f(z) = \frac{(-1)^{N+1}N!}{2\pi jh}\int_{C_r}\frac{f(w)}{\left(\frac{z-w}{h}\right)_{N+1}}\mathrm{d}w \tag{3.4}$$

Attending to the relation between the Pochhammer symbol and the Gamma function:

$$\Gamma(z+n) = (z)_n \Gamma(z) \tag{3.5}$$

we can write:

$$\Delta_d^N f(z) = \frac{N!}{2\pi jh} \int_{C_d} f(w) \frac{\Gamma\left(\frac{w-z}{h}\right)}{\Gamma\left(\frac{w-z}{h}+N+1\right)} dw \tag{3.6}$$

and

$$\Delta_r^N f(z) = \frac{(-1)^{N+1} N!}{2\pi jh} \int_{C_r} f(w) \frac{\Gamma\left(\frac{z-w}{h}\right)}{\Gamma\left(\frac{z-w}{h}+N+1\right)} dw \tag{3.7}$$

This is correct and is coherent with the difference definitions, because the Gamma function $\Gamma(z)$ has poles at the negative integers ($z = -n$). The corresponding residues are equal to $(-1)^n/n!$. Although both the Gamma functions have infinite poles, outside the contour they cancel out and the integrand is analytic. We should also remark that the direct and reverse differences are not equal.

3.2.2 Fractional Order

Consider the fractional order differences defined in (2.13) and (2.14). It is not hard to see that we are in presence of a situation similar to the positive integer case, excepting the fact of having infinite poles. So we have to use an integration path that encircles all the poles. This can be done with a U shaped contour like those shown in Fig. 3.2. We use (3.6) and (3.7) with the suitable adaptations, obtaining:

$$\Delta_d^\alpha f(z) = \frac{\Gamma(\varepsilon+1)}{2\pi jh} \int_{C_d} f(w) \frac{\Gamma\left(\frac{w-z}{h}\right)}{\Gamma\left(\frac{w-z}{h}+\alpha+1\right)} dw \tag{3.8}$$

and

$$\Delta_r^\alpha f(z) = \frac{(-1)^{\alpha+1} \Gamma(\alpha+1)}{2\pi jh} \int_{C_r} f(w) \frac{\Gamma\left(\frac{z-w}{h}\right)}{\Gamma\left(\frac{z-w}{h}+\alpha+1\right)} dw \tag{3.9}$$

Remark that one turns into the other with the substitution h → −h. We can use the residue theorem to confirm the correctness of the above formulae.

Fig. 3.2 Integration paths and poles for the integral representation of fractional order differences

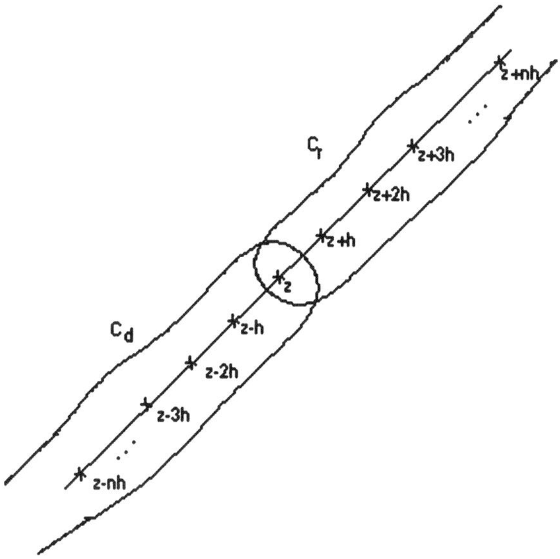

3.2.3 *Two Properties*

In the following, we shall be concerned with the fractional order case. We will consider the direct case. The other is similar.

3.2.3.1 Repeated differencing

We are going to study the effect of a sequential application of the difference operator Δ. We have

$$\Delta_d^{\beta}[\Delta_d^{\alpha} f(z)] = \frac{\Gamma(\beta+1)\Gamma(\alpha+1)}{(2\pi jh)^2} \int_{C_d} \int_{C_d} f(w) \frac{\Gamma\left(\frac{w-s}{h}\right)}{\Gamma\left(\frac{w-s}{h}+\alpha+1\right)} \mathrm{d}w \frac{\Gamma\left(\frac{s-z}{h}\right)}{\Gamma\left(\frac{s-z}{h}+\beta+1\right)} \mathrm{d}s \tag{3.10}$$

Permuting the integrations, we obtain

$$\Delta_d^{\beta}[\Delta_d^{\alpha} f(z)] = \frac{\Gamma(\beta+1)\Gamma(\alpha+1)}{(2\pi jh)^2} \int_{C_d} f(w) \int_{C_d} \frac{\Gamma\left(\frac{w-s}{h}\right)}{\Gamma\left(\frac{w-s}{h}+\alpha+1\right)} \frac{\Gamma\left(\frac{s-z}{h}\right)}{\Gamma\left(\frac{s-z}{h}+\beta+1\right)} \mathrm{d}s\, \mathrm{d}w \tag{3.11}$$

By the residue theorem

$$\begin{aligned}
&\frac{\Gamma(\beta+1)}{2\pi jh}\int_{C_d}\frac{\Gamma\left(\frac{w-s}{h}\right)}{\Gamma\left(\frac{w-s}{h}+\alpha+1\right)}\frac{\Gamma\left(\frac{s-z}{h}\right)}{\Gamma\left(\frac{s-z}{h}+\beta+1\right)}\mathrm{d}s\\
&=\frac{1}{h}\sum_{n=0}^{\infty}\frac{(-1)^n}{n!}\frac{\Gamma(\beta+1)\cdot\Gamma\left(\frac{w-z}{h}+n\right)}{\Gamma\left(\frac{w-z}{h}+\alpha+1+n\right)\Gamma(\beta-n+1)}\\
&=\frac{1}{h}\frac{\Gamma\left(\frac{w-z}{h}\right)}{\Gamma\left(\frac{w-z}{h}+\alpha+1\right)}\sum_{n=0}^{\infty}\frac{\left(\frac{w-z}{h}\right)_n(-\beta)_n}{\left(\frac{w-z}{h}+\alpha+1\right)_n}\\
&=\frac{1}{h}\frac{\Gamma\left(\frac{w-z}{h}\right)}{\Gamma\left(\frac{w-z}{h}+\alpha+1\right)}{}_2F_1\left(\frac{w-z}{h};-\beta;\frac{w-z}{h}+\alpha+1;1\right)
\end{aligned}\tag{3.12}$$

where ${}_2F_1$ is the Gauss hypergeometric function. If $\alpha+\beta+1>0$, we have:

$$\frac{\Gamma(\beta+1)}{2\pi jh}\int_{C_d}\frac{\Gamma\left(\frac{w-s}{h}\right)}{\Gamma\left(\frac{w-s}{h}+\alpha+1\right)}\frac{\Gamma\left(\frac{s-z}{h}\right)}{\Gamma\left(\frac{s-z}{h}+\beta+1\right)}\mathrm{d}s=\frac{1}{h}\frac{\Gamma\left(\frac{w-z}{h}\right)\Gamma(\alpha+\beta+1)}{\Gamma\left(\frac{w-z}{h}+\alpha+\beta+1\right)\Gamma(\alpha+1)}\tag{3.13}$$

leading to the conclusion that:

$$\Delta_d^{\beta}[\Delta_d^{\alpha}f(z)]=\frac{\Gamma(\alpha+\beta+1)}{2\pi jh}\int_{C_d}f(w)\frac{\Gamma\left(\frac{w-z}{h}\right)}{\Gamma\left(\frac{w-z}{h}+\alpha+\beta+1\right)}\mathrm{d}w\tag{3.14}$$

and

$$\Delta_d^{\beta}[\Delta_d^{\alpha}f(z)]=\Delta_d^{\alpha+\beta}f(z)\tag{3.15}$$

provided that $\alpha+\beta+1>0$. It is not difficult to see that the above operation is commutative. The condition $\alpha+\beta+1>0$ is restrictive, since it may happen that we cannot have $\beta\le-\alpha-1$. However, we must remark that (3.8) and (3.9) are valid for every $\alpha\in\mathbf{R}$. The same happens with $(\alpha+\beta)$ in (3.15). This means that we can use (3.15) with every pair $(\alpha,\beta)\in\mathbf{R}$. It should be stressed that, at least in principle, we must not mix the two differences, because they use different integration paths. If any way we decide to do it, we have to use a doubly opened integration path. The result seem not to have any interest here. Later we will use such a path when dealing with the centred derivatives.

3.2.3.2 Inversion

Putting $\alpha=-\beta$ into (3.15), we obtain:

$$\begin{aligned}
\Delta_d^{-\alpha}[\Delta_d^{\alpha}f(z)]&=\Delta_d^{\alpha}[\Delta_d^{-\alpha}f(z)]\\
&=\frac{1}{2\pi j}\int_{C_d}f(w)\frac{1}{w-z}\mathrm{d}w=f(z)
\end{aligned}\tag{3.16}$$

as we would expect. So the operation of differencing is invertible. This means that we can write:

$$f(z) = \frac{\Gamma(\alpha+1)}{2\pi jh} \int\limits_{C_d} \Delta_d^\alpha f(w) \frac{\Gamma\left(\frac{w-z}{h}\right)}{\Gamma\left(\frac{w-z}{h} - \alpha + 1\right)} \mathrm{d}w \tag{3.17}$$

in the direct case. In the reverse case, we will have:

$$f(z) = \frac{\Gamma(\alpha+1)}{2\pi jh} \int\limits_{C_d} \Delta_r^\alpha f(w) \frac{\Gamma\left(\frac{z-w}{h}\right)}{\Gamma\left(\frac{z-w}{h} - \alpha + 1\right)} \mathrm{d}w \tag{3.18}$$

according to (3.9).

3.3 Obtaining the Generalized Cauchy Formula

The ratio of two gamma functions $\frac{\Gamma(s+a)}{\Gamma(s+b)}$ has an interesting expansion [5]:

$$\frac{\Gamma(s+a)}{\Gamma(s+b)} = s^{a-b}\left[1 + \sum_1^N C_k s^{-k} + O\left(s^{-N-1}\right)\right] \tag{3.19}$$

as $|s| \to \infty$, uniformly in every sector that excludes the negative real half-axis. The coefficients in the series can be expressed in terms of Bernoulli polynomials. Their knowledge is not important here.

Consider (3.8) and (3.9) again. Let $|h| < \varepsilon \in R$, where ε is a small number. This allows us to write:

$$\Delta_d^\alpha f(z) = \frac{\Gamma(\alpha+1)}{2\pi jh} \int\limits_{C_d} f(w) \frac{1}{\left(\frac{w-z}{h}\right)^{\alpha+1}} \mathrm{d}w + g_1(h) \tag{3.20}$$

and

$$\Delta_r^\alpha f(z) = \frac{(-1)^{\alpha+1}\Gamma(\alpha+1)}{2\pi jh} \int\limits_{C_r} f(w) \frac{1}{\left(\frac{z-w}{h}\right)^{\alpha+1}} \mathrm{d}w + g_2(h) \tag{3.21}$$

where C_d and C_r are the contours represented in Fig. 3.2. The $g_1(h)$ and $g_2(h)$ terms are proportional to $h^{\alpha+2}$. So, the fractional incremental ratio are, aside terms proportional to h, given by:

$$\frac{\Delta_d^\alpha f(z)}{h^\alpha} = \frac{\Gamma(\alpha+1)}{2\pi j} \int\limits_{C_d} f(w) \frac{1}{(w-z)^{\alpha+1}} \mathrm{d}w \tag{3.22}$$

and

$$\frac{\Delta_r^\alpha f(z)}{h^\alpha} = \frac{\Gamma(\alpha+1)}{2\pi j} \int\limits_{C_r} f(w) \frac{1}{(w-z)^{\alpha+1}} \mathrm{d}w \tag{3.23}$$

Allowing $h \to 0$, we obtain the direct and reverse generalised Cauchy derivatives:

$$D_d^\alpha f(z) = \frac{\Gamma(\alpha+1)}{2\pi j} \int\limits_{C_d} f(w) \frac{1}{(w-z)^{\alpha+1}} \mathrm{d}w \tag{3.24}$$

and

$$D_r^\alpha f(z) = \frac{\Gamma(\alpha+1)}{2\pi j} \int\limits_{C_r} f(w) \frac{1}{(w-z)^{\alpha+1}} \mathrm{d}w \tag{3.25}$$

If $\alpha = N$, both the derivatives are equal and coincide with the usual Cauchy definition. In the fractional case we have different solutions, since we are using a different integration path. Remark that (3.24) and (3.25) are formally the same. They differ only in the integration path. This means that we can use a general procedure as we did on Chap. 2.

Definition 3.1 We define the generalised Cauchy derivative by

$$D_\theta^\alpha f(z) = \frac{\Gamma(\alpha+1)}{2\pi j} \int\limits_{C_\theta} f(w) \frac{1}{(w-z)^{\alpha+1}} \mathrm{d}w \tag{3.26}$$

where C_θ is any U-shaped path encircling the branch cut line and making an angle $\theta + \pi$ with the real positive half axis. As we will see next, the particular cases $\theta = 0$ or $\theta = \pi$ lead to new forms of representing the forward and backward derivatives.

3.4 Analysis of Cauchy Formula

3.4.1 General Formulation

Consider the generalised Cauchy formula (3.26) and rewrite it in a more convenient format obtained by a simple translation:

$$D_\theta^\alpha f(z) = \frac{\Gamma(\alpha+1)}{2\pi j} \int\limits_{C} f(w+z) \frac{1}{w^{\alpha+1}} \mathrm{d}w \tag{3.27}$$

Here we will choose C as a special integration path: the Hankel contour represented in Fig. 3.3. It is constituted by two straight lines and a small circle. We assume that it surrounds the selected branch cut line. This is described by $x \cdot e^{j\theta}$, with $x \in \mathbf{R}^+$ and $\theta \in (0,2\pi[$. The circle has a radius equal to ρ small enough to allow it to stay inside the region of analyticity of $f(z)$. If α is a negative integer, the integral along the circle is zero and we are led to the well known repeated integration formula [6–8] as we will see later. In the general α case we need the two terms. Let us decompose the above integral using the Hankel contour. For reducing steps, we will assume already that the straight lines are infinitely near. We have, then:

$$D_{\theta}^{\alpha} f(z) = \frac{\Gamma(\alpha+1)}{2\pi j} \left[\int_{C_1} + \int_{C_2} + \int_{C_3} \right] f(w+z) \frac{1}{w^{\alpha+1}} \mathrm{d}w \tag{3.28}$$

Over C_1 we have $w = x \cdot e^{j(\theta-\pi)}$, while over C_3 we have $w = x \cdot e^{j(\theta+\pi)}$, with $x \in \mathbf{R}^+$, over C_2 we have $w = \rho e^{j\varphi}$, with $\varphi \in]\theta - \pi, \theta+\pi[$. We can write, at last:

$$D_{\theta}^{\alpha} f(z) = \frac{\Gamma(\alpha+1)}{2\pi j} \left[\int_{\infty}^{\rho} f(x \cdot e^{j(\theta-\pi)} + z) \frac{e^{-j\alpha(\theta-\pi)}}{x^{\alpha+1}} \mathrm{d}x + \int_{\rho}^{\infty} f(x \cdot e^{j(\theta+\pi)} + z) \frac{e^{-j\alpha(\theta+\pi)}}{x^{\alpha+1}} \mathrm{d}x \right]$$
$$+ \frac{\Gamma(\alpha+1)}{2\pi j} \frac{1}{\rho^{\alpha}} \int_{\theta-\pi}^{\theta+\pi} f(\rho \cdot e^{j\varphi} + z) e^{-j\alpha\varphi} j \, \mathrm{d}\varphi \tag{3.29}$$

For the first term, we have:

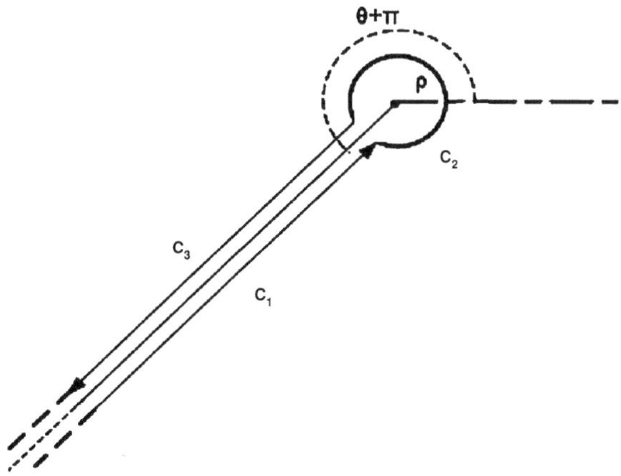

Fig. 3.3 The Hankel contour used in computing the derivative defined in Eq. 3.27

$$\int_{\infty}^{\rho} f(x \cdot e^{j(\theta-\pi)} + z)\frac{e^{-j\alpha(\theta-\pi)}}{x^{\alpha+1}}dx + \int_{\rho}^{\infty} f(x \cdot e^{j(\theta+\pi)} + z)\frac{e^{-j\alpha(\theta+\pi)}}{x^{\alpha+1}}dx$$
$$= [-e^{-j\alpha(\theta-\pi)} + e^{-j\alpha(\theta+\pi)}]\int_{\rho}^{\infty} f(-x \cdot e^{j\theta} + z)\frac{1}{x^{\alpha+1}}dx$$
$$= -e^{-j\alpha\theta} \cdot [e^{j\pi\alpha} - e^{-j\pi\alpha}]\int_{\rho}^{\infty} f(-x \cdot e^{j\theta} + z)\frac{1}{x^{\alpha+1}}dx$$
$$= -e^{-j\alpha\theta} \cdot 2j \cdot \sin(\alpha\pi)\int_{\rho}^{\infty} f(-x \cdot e^{j\theta} + z)\frac{1}{x^{\alpha+1}}dx \tag{3.30}$$

where we assumed that $f(-x{\cdot}e^{j(\theta-\pi)} + z) = f(-x{\cdot}e^{j(\theta+\pi)} + z)$, because $f(z)$ is analytic.

For the second term, we begin by noting that the analyticity of the function $f(z)$ allows us to write:

$$f(-x \cdot e^{j\theta} + z) = \sum_{0}^{\infty}\frac{f^{(n)}(z)}{n!}(-1)^n x^n e^{jn\theta} \tag{3.31}$$

for $x < r \in \mathbf{R}^+$. We have, then:

$$j\frac{1}{\rho^\alpha}\int_{\theta-\pi}^{\theta+\pi} f(\rho \cdot e^{j\varphi} + z)e^{-j\alpha\varphi}\, d\varphi = j\frac{1}{\rho^\alpha}\sum_{0}^{\infty}\frac{f^{(n)}(z)}{n!}\rho^n\int_{\theta-\pi}^{\theta+\pi} e^{j(n-\alpha)\varphi}d\varphi \tag{3.32}$$

Performing the integration, we have:

$$\begin{aligned} j\frac{1}{\rho^\alpha}\int_{\theta-\pi}^{\theta+\pi} f(\rho \cdot e^{j\varphi} + z)e^{-j\alpha\varphi}d\varphi &= -j\sum_{0}^{\infty}\frac{f^{(n)}(z)}{n!}\rho^{n-\alpha}e^{j(n-\alpha)\theta}\frac{2 \cdot \sin[(n-\alpha)\pi]}{(n-\alpha)} \\ &= 2j \cdot e^{-j\alpha\theta}\sin(\alpha\pi)\sum_{0}^{\infty}\frac{f^{(n)}(z)}{n!}\frac{e^{jn\theta}(-1)^n\rho^{n-\alpha}}{(n-\alpha)} \end{aligned} \tag{3.33}$$

But the summation in the last expression can be written in another interesting format:

$$\sum_0^\infty \frac{f^{(n)}(z)}{n!}\frac{e^{jn\theta}(-1)^n \rho^{n-\alpha}}{(n-\alpha)}$$
$$= \left[-\sum_0^N \frac{f^{(n)}(z)}{n!}(-1)^n e^{jn\theta} \int_\rho^\infty x^{n-\alpha-1}dx + \sum_{N+1}^\infty \frac{f^{(n)}(z)}{n!}\frac{(-1^n)e^{jn\theta}\rho^{n-\alpha}}{(n-\alpha)} \right]$$

where $N = \lfloor \alpha \rfloor$.[2] Substituting it in (3.33) and joining to (3.30) we can write:

$$D_\theta^\alpha f(z) = K \cdot \int_\rho^\infty \frac{\left[f(-x \cdot e^{j\theta} + z) - \sum_0^N \frac{f^{(n)}(z)}{n!}(-1)^n e^{jn\theta} x^n \right]}{x^{\alpha+1}} dx$$
$$- K \sum_{N+1}^\infty \frac{f^{(n)}(z)}{n!}(-1)^n e^{jn\theta} \frac{\rho^{n-\alpha}}{(n-\alpha)} \tag{3.34}$$

If $\alpha < 0$, we make the inner summation equal to zero. Using the reflection formula of the gamma function

$$\frac{1}{\Gamma(\beta)\Gamma(1-\beta)} = \frac{\sin(\pi\beta)}{\pi}$$

we obtain for K

$$K = -\frac{\Gamma(\alpha+1)e^{-j\theta\alpha}}{\pi}\sin(\alpha\pi) = \frac{e^{-j\theta\alpha}}{\Gamma(-\alpha)} \tag{3.35}$$

Now let ρ go to zero in (3.34). The second term on the right hand side goes to zero and we obtain:

$$D_\theta^\alpha f(z) = \frac{e^{-j\theta\alpha}}{\Gamma(-\alpha)} \int_0^\infty \frac{\left[f(-x \cdot e^{j\theta} + z) - \sum_0^N \frac{f^{(n)}(z)}{n!}(-1)^n e^{jn\theta} x^n \right]}{x^{\alpha+1}} dx \tag{3.36}$$

that is valid for any $\alpha \in \mathbf{R}$.

It is interesting to remark that (3.36) is nothing else than a generalisation of the "pseudo-function" notion [9, 10], but valid for an analytic function in a non compact region of the complex plane. On the other hand, *we did not have to reject any infinite part* as Hadamard did. Relation (3.36) represents a regularised fractional derivative that has some similarities with the Marchaud derivative [5]: for $0 < \alpha < 1$, they are equal.

If one puts $w = x{\cdot}e^{j\theta}$, we can write:

---

[2] $\lfloor \alpha \rfloor$ means "the least integer less than or equal to α".

$$D_{\theta}^{\alpha}f(z) = \frac{1}{\Gamma(-\alpha)}e^{-j\theta\alpha}\int\limits_{\gamma_\theta}\frac{\left[f(z-w) - \sum_0^N \frac{f^{(n)}(z)(-1)^n}{n!}w^n\right]}{w^{\alpha+1}}\mathrm{d}w \tag{3.37}$$

where γ_θ is a half straight line starting at $w = 0$ and making an angle θ with the positive real axis. As we can conclude there are infinite ways of computing the derivative of a given function: these are defined by the chosen branch cut lines. However, this does not mean that we have infinite different derivatives. It is not hard to see that all the branch cut lines belonging to a given region of analyticity of the function are equivalent and lead to the same result unless the integral may be divergent if the function increases without bound.

3.5 Examples

3.5.1 The Exponential Function

To illustrate the previous assertions we are going to consider the case of the exponential function.

Let $f(z) = e^{az}$, with $a \in \mathbf{R}$. Inserting it into (3.36), it comes:

$$D_{\theta}^{\alpha}f(z) = \frac{1}{\Gamma(-\alpha)}e^{-j\theta\alpha}e^{az}\int\limits_0^\infty\frac{\left[e^{-ax\cdot e^{j\theta}} - \sum_0^N \frac{a^n}{n!}e^{jn\theta}(-1)^n x^n\right]}{x^{\alpha+1}}\mathrm{d}x$$

With a variable change $\tau = axe^{j\theta}$, the above equation gives:

$$D_{\theta}^{\alpha}f(z) = \frac{1}{\Gamma(-\alpha)}a^\alpha e^{az}\int\limits_0^{\infty\cdot ae^{j\theta}}\frac{\left[e^{-\tau} - \sum_0^N \frac{(-1)^n}{n!}\tau^n\right]}{\tau^{\alpha+1}}\mathrm{d}\tau \tag{3.38}$$

where the integration path is half straight line that forms an angle equal to θ with the positive real axis, in agreement with (3.36). The integral in (3.38) is almost the generalised Gamma function definition [11, 12] and is a generalisation of Euler integral representation for the gamma function. But this requires integration along the positive real axis. However, the integration can be done along any ray with an angle in the interval $[0, \pi/2)$ [13]. To obtain convergence of this integral we must have $Re(ae^{j\theta}) > 0$. This means that **a** must necessarily be positive. This is coherent with what was said in Chap. 2: the forward derivative $\{|\theta| \in [0, \pi/2)\}$ of an exponential exists only if the function behaves like "right" function going to zero when z goes to $-\infty$. Returning to the above integral, we can write:

$$D_{\theta}^{\alpha}f(z) = \frac{1}{\Gamma(-\alpha)}a^\alpha e^{az}\int\limits_0^{\infty}\frac{\left[e^{-\tau} - \sum_0^N \frac{(-1^n)}{n!}\tau^n\right]}{(\tau^{\alpha+1})}\mathrm{d}\tau \tag{3.39}$$

The integral defines the value of the gamma function $\Gamma(-\alpha)$. In fact [11, 12] we have

$$\Gamma(z) = \int_0^\infty \tau^{z-1}\left[\mathrm{e}^{-\tau} - \sum_0^N \frac{(-1)^n}{n!}\tau^n\right]\mathrm{d}\tau \tag{3.40}$$

if we maintain the convention made before: when $z < 0$ the summation is zero. We obtain then:

$$D_\theta^\alpha[\mathrm{e}^{az}] = a^\alpha \mathrm{e}^{az} \tag{3.41}$$

as expected. In the particular limiting case, $a \to 0$, we obtain $D_\theta^\alpha 1 = 0$ provided that $\alpha > 0$. If $\alpha < 0$, the limit is infinite. The Grünwald–Letnikov definition allowed us to obtain the same conclusions as seen. Now, consider the case where $a < 0$. To obtain convergence in (3.38) we must have $\theta \in [\pi/2, 3\pi/2)$. This means that the exponential must go to zero when z goes to $+\infty$. It is what we called a "left" function. The derivative is also expressed by (3.41) but the branch cut line to define the power is now a half straight line in the right half complex plane: in particular the positive real half axis. This is the same problem we found in Sect. 2.7.3. We conclude then that (3.41) is the result given by the forward derivative if $a > 0$ and the one given by the backward derivative if $a < 0$. This has very important consequences. By these facts we must be careful in using (3.41). In fact, in a first glance, we could be lead to use it to compute the derivatives of functions like $\sin(z)$, $\cos(z)$, $\sinh(z)$ and $\cosh(z)$. But if we have in mind our reasoning we can conclude immediately that those functions do not have finite derivatives if $z \in \mathbf{C}$.

In fact they use simultaneously the exponentials e^z and e^{-z} which derivatives cannot exist simultaneously, as we just saw. However, we can conclude that functions expressed by Dirichlet series $f(t) = \sum_0^\infty a_k \mathrm{e}^{\lambda_k t}$ with all the $Re(\lambda_k)$ positive or all negative have finite derivatives given by $f^{(\alpha)}(t) = \sum_0^\infty a_k(\lambda_k)^\alpha \mathrm{e}^{\lambda_k t}$. In particular functions with Laplace transform with region of convergence in the right or left half planes have fractional derivatives.

Another interesting case is the cisoid $f(t) = \mathrm{e}^{j\omega t}$, $\omega \in \mathrm{R}^+$. Inserting it into (3.36) again, it comes:

$$D_\theta^\alpha f(t) = \frac{1}{\Gamma(-\alpha)}\mathrm{e}^{-j\theta\alpha}\mathrm{e}^{j\omega t}\int_0^\infty \frac{\left[\mathrm{e}^{j\omega x.\mathrm{e}^{j\theta}} - \sum_0^N \frac{(j\omega)^n}{n!}\mathrm{e}^{jn\theta}x^n\right]}{x^{\alpha+1}}\mathrm{d}x \tag{3.42}$$

With $\theta = \pi/2$, $j\omega\mathrm{e}^{j\theta} = -\omega$ and we obtain easily:

$$D_f^\alpha f(t) = (j\omega)^\alpha \mathrm{e}^{j\omega t}$$

It is not difficult to see that (3.42) remains valid if $\omega < 0$ provided that we remember that the branch cut line is the negative real half axis. We only have to put $\theta = -\pi/2$

$$D^{\alpha}_{f}f(t) = (-j\omega)^{\alpha}\mathrm{e}^{-j\omega t}$$

We can conclude then that:

$$D^{\alpha}_{f}\cos(\omega t) = \omega^{\alpha}\cos(\omega t + \alpha\pi/2) \tag{3.43}$$

This procedure corresponds to extend the validity of the forward derivative and agrees with the results we presented in Sect. 2.7.4. For $\sin(\omega t)$, the procedure is similar leading to

$$D^{\alpha}_{f}\sin(\omega t) = \omega^{\alpha}\sin(\omega t + \alpha\pi/2) \tag{3.44}$$

When $\alpha = 1$, we recover the usual formulae. The backward case would lead to the results obtained in Sect. 2.7.3. We will not consider it again.

3.5.2 The Power Function

Let $f(z) = z^{\beta}$, with $\beta \in R$. If $\beta > \alpha$, we will show that $D^{\alpha}[z^{\beta}]$ defined for every $z \in \mathbf{C}$ does not exist, unless α is a positive integer, because the integral in (3.36) is divergent for every $\theta \in [-\pi, \pi)$. This has an important consequence: *we cannot compute the derivative of a given function by using its Taylor series and computing the derivative term by term.*

Let us see what happens for non integer values of α. The branch cut line needed for the definition of the function must be chosen to be outside the integration region. This is equivalent to say that the two branch cut lines cannot intersect. To use (3.36) we compute the successive integer order derivatives of this function that are given by:

$$D^{n}z^{\beta} = (-1)^{n}(-\beta)_{n}z^{\beta-n} \tag{3.45}$$

Now, we have:

$$D^{\alpha}_{\theta}z^{\beta} = \frac{\mathrm{e}^{-j\theta\alpha}}{\Gamma(-\alpha)}\int\limits_{0}^{\infty}\frac{\left[(-x\cdot\mathrm{e}^{j\theta}+z)^{\beta} - \sum_{0}^{N}\frac{(-1)^{n}(-\beta)_{n}z^{\beta-n}}{n!}\mathrm{e}^{jn\theta}x^{n}\right]}{x^{\alpha+1}}\mathrm{d}x \tag{3.46}$$

With a substitution $\tau = x\cdot\mathrm{e}^{j\theta}/z$, we obtain:

$$D^{\alpha}_{\theta}z^{\beta} = \frac{1}{\Gamma(-\alpha)}z^{\beta-\alpha}\int\limits_{0}^{\infty\,\mathrm{e}^{j\theta}/z}\frac{\left[(1-\tau)^{\beta} - \sum_{0}^{N}\frac{(-1)^{n}(-\beta)_{n}}{n!}\tau^{n}\right]}{\tau^{\alpha+1}}\mathrm{d}\tau \tag{3.47}$$

To become simpler the analysis let us assume that $\theta = 0$ and $z \in \mathbf{R}^+$. We obtain:

$$D_f^\alpha z^\beta = \frac{1}{\Gamma(-\alpha)} z^{\beta-\alpha} \int_0^\infty \frac{\left[(1-\tau)^\beta - \sum_0^N \frac{(-1)^n (-\beta)_n}{n!} \tau^n\right]}{\tau^{\alpha+1}} \mathrm{d}\tau \tag{3.48}$$

Let us decompose the integral

$$\int_0^\infty \frac{\left[(1-\tau)^\beta - \sum_0^N \frac{(-1)^n (-\beta)_n}{n!} \tau^n\right]}{\tau^{\alpha+1}} \mathrm{d}\tau = \int_0^1 \frac{\left[(1-\tau)^\beta - \sum_0^N \frac{(-1)^n (-\beta)_n}{n!} \tau^n\right]}{\tau^{\alpha+1}} \mathrm{d}\tau + \int_1^\infty \frac{\left[(1-\tau)^\beta - \sum_0^N \frac{(-1)^n (-\beta)_n}{n!} \tau^n\right]}{\tau^{\alpha+1}} \mathrm{d}\tau$$

As shown in Ortigueira [14], the first integral is a generalised version of the Beta function $B(-\alpha, \beta+1)$ valid for $\alpha \in \mathbf{R}$ and $\beta > -1$. But the second is divergent. We conclude that the power function defined in $\mathbf{C}$ does not have fractional derivatives.

3.5.3 The Derivatives of Real Functions

As we are mainly interested in real variable functions we are going to obtain the formulae suitable for this case. Now, we only have two hypotheses: $\theta = 0$ or $\theta = \pi$.

3.5.3.1 $\theta = 0$: Forward Derivative

This corresponds to choosing the real negative half axis as branch cut line. Substituting $\theta = 0$ into (3.36), we have:

$$D_f^\alpha f(z) = \frac{1}{\Gamma(-\alpha)} \int_0^\infty \frac{\left[f(z-x) - \sum_0^N \frac{f^{(n)}(z)}{n!} (-x)^n\right]}{x^{\alpha+1}} \mathrm{d}x \tag{3.49}$$

As this integral uses the left hand values of the function, we will call this forward or direct derivative again in agreement with Sect. 2.3.

3.5.3.2 $\theta = \pi$: Backward Derivative

This corresponds to choosing the real positive half axis as branch cut line. Substituting $\theta = \pi$ into (3.36) and performing the change $x \to -x$, we have:

$$D_b^\alpha f(z) = \frac{-j\pi\alpha}{\Gamma(-\alpha)} \int_0^\infty \frac{\left[f(x+z) - \sum_0^N \frac{f^{(n)}(z)}{n!} x^n\right]}{x^{\alpha+1}} \mathrm{d}x \tag{3.50}$$

As this integral uses the right hand values of the function, we will call this backward or reverse derivative in agreement with Sect. 2.3 again.

3.5.4 Derivatives of Some Causal Functions

We are going to study the causal power function and exponential. Although we could do it using the LT as we will see in the next section, we are going to do it here using the relation (3.49). Let $f(t) = t^\beta u(t)$. As seen above:

$$D^n t^\beta u(t) = (-1)^n (-\beta)_n t^{\beta-n} u(t)$$

that inserted in (3.49) gives

$$D_f^\alpha f(t) = t^\beta u(t) \frac{1}{\Gamma(-\alpha)} \int_0^t \frac{\left[(1-x/t)^\beta - \sum_0^N \frac{(-1)^n(-\beta)_n t^{-n}}{n!} (-x)^n\right]}{x^{\alpha+1}} \mathrm{d}x$$

that is converted in the next expression through the substitution $\tau = x/t$

$$D_f^\alpha t^\beta = t^{\beta-\alpha} u(t) \frac{1}{\Gamma(-\alpha)} \int_0^1 \frac{\left[(1-\tau)^\beta - \sum_0^N \frac{(-1)^n(-\beta)_n}{n!} (-\tau)^n\right]}{\tau^{\alpha+1}} \mathrm{d}\tau.$$

The above integral is a representation of the beta function, $B(-\alpha, \beta + 1)$, for $\beta > -1$ {see [14]}. But

$$B(-\alpha, \beta+1) = \frac{\Gamma(-\alpha)\Gamma(\beta+1)}{\Gamma(b-\alpha+1)}$$

and then

$$D_f^\alpha t^\beta = \frac{\Gamma(\beta+1)}{\Gamma(\beta-\alpha+1)} t^{\beta-\alpha} u(t) \tag{3.51}$$

that coincides with the result obtained in (2.76).

Now let us try the exponential function, $f(t) = e^{at}u(t)$. Inserting in (3.49) we obtain

$$D_f^\alpha e^{at} u(t) = -e^{at} u(t) \frac{1}{\Gamma(-\alpha)} \int_0^t \frac{\left[\sum_{N+1}^\infty \frac{(-a)^n x^n}{n!}\right]}{x^{\alpha+1}} \mathrm{d}x$$

Assuming that the series converges uniformly, we get easily

$$D_f^{\alpha} e^{at} u(t) = -e^{at} u(t) \frac{1}{\Gamma(-\alpha)} \left[\sum_{N+1}^{\infty} \frac{(-a)^n t^{n-\alpha}}{n!(n-\alpha)} \right] \tag{3.52}$$

Alternatively we can use the causal part of the McLaurin series and compute the derivative of each term. This is not a contradiction with our affirmation because the terms of the series are causal powers.

3.6 Derivatives of Functions with Laplace Transform

Consider now the special class of functions with Laplace Transform. Let $f(t)$ be such a function and $F(s)$ its LT, with a suitable region of convergence, R_c. This means that we can write

$$f(t) = \frac{1}{2\pi j} \int_{a-j\infty}^{a+j\infty} F(s) e^{st} \mathrm{d}s \tag{3.53}$$

where a is a real number inside the region of convergence. Inserting (3.53) inside (3.49) and permuting the integration symbols, we obtain:

$$D_f^{\alpha} f(z) = \frac{1}{2\pi j \Gamma(-\alpha)} \int_{a-j\infty}^{a+j\infty} F(s) e^{sz} \int_0^{\infty} \frac{\left[e^{-sx} - \sum_0^N \frac{(sx)^n}{n!} \right]}{x^{\alpha+1}} \mathrm{d}x \, \mathrm{d}s \tag{3.54}$$

Now we are going to use the results presented above in Sect. 3.5. If $Re(s) > 0$, the inner integral is equal to $\Gamma(-\alpha) \cdot s^{\alpha}$, if $Re(s) < 0$ it is divergent. We conclude that:

$$LT[D_f^{\alpha} f(t)] = s^{\alpha} F(s) \quad \text{for} \quad Re(s) > 0 \tag{3.55}$$

a well known result.

Now, insert (3.53) inside (3.50) and permute again the integration symbols to obtain

$$D_b^{\alpha} f(z) = \frac{e^{-j\pi\alpha}}{2\pi j \Gamma(-\alpha)} \int_{a-j\infty}^{a+j\infty} F(s) e^{sz} \int_0^{\infty} \frac{\left[e^{sx} - \sum_0^N \frac{(sx)^n}{n!} \right]}{x^{\alpha+1}} \mathrm{d}x \, \mathrm{d}s \tag{3.56}$$

If $Re(s) < 0$ and considering the result obtained in Sect. 3.5 {see Sect. 2.7 also} the inner integral is equal to $\Gamma(-\alpha) \cdot s^{\alpha}$, if $Re(s) > 0$ it is divergent. We conclude that:

$$LT[D_b^{\alpha} f(t)] = s^{\alpha} F(s) \quad \text{for} \quad Re(s) < 0 \tag{3.57}$$

We confirmed the results we obtained in Chap. 2 stating the enlargement the applicability of the well known property of the Laplace transform of the derivative. The presence of the factor $e^{-j\pi\alpha}$ may seem strange but is a consequence of assuming that $H(s) = s^{\alpha}$ is the common expression for the transfer function of the causal and anti-causal differintegrator. We must be careful because in current literature that factor has been removed and the resulting derivative is called "right" derivative. According to the development we did that factor must be retained. It is interesting to refer that this was already none by Liouville.

3.7 Generalized Caputo and Riemann–Liouville Derivatives for Analytic Functions

The most known and popular fractional derivatives are almost surely the Riemann–Liouville (RL) and the Caputo (C) derivatives [5, 8, 15]. Without considering the reserves put before [14], we are going to face two related questions:

- Can we formulate those derivatives in the complex plane?
- Is there a coherent relation between those derivatives and the incremental ratio based Grünwald–Letnikov (GL)?

As expected attending to what we wrote in Chap. 2 about these derivatives, the answers for those questions are positive. We proceed by constructing formulations in the complex plane obtained from the GL as we did in Sect. 3.5.

3.7.1 RL and C Derivatives in the Complex Plane

As we showed in Sect. 2.6, the generalised GL derivative verifies

$$D_{\theta}^{\alpha}\left[D_{\theta}^{\beta}f(t)\right] = D_{\theta}^{\beta}\left[D_{\theta}^{\alpha}f(t)\right] = D_{\theta}^{\alpha+\beta}f(t) \tag{3.58}$$

provided that both derivatives (of orders α and β) exist. This is what we called before the semi group property. This is important and not enjoyed by other derivatives.

In particular we can put $\beta = n \in \mathbf{Z}^{+}$ and $\varepsilon = n - \alpha > 0$ and we are led to the following expressions

$$D_{\theta}^{\alpha}f(z) = D^{n}\left[e^{j\theta\varepsilon}\lim_{|h|\to 0}\sum_{k=0}^{\infty}(-1)^{k}\binom{-\varepsilon}{k}f(z-kh)|h|^{\varepsilon}\right] \tag{3.59}$$

and

$$D_{\theta}^{\alpha}f(z) = e^{j\theta\varepsilon}\lim_{|h|\to 0}\sum_{k=0}^{\infty}(-1)^{k}\binom{-\varepsilon}{k}f^{(n)}(z-kh)|h|^{\varepsilon} \tag{3.60}$$

that we can call mix GL-RL and GL-C.

According to what we showed in Sect. 3.3 the GL derivative leads to the generalised Cauchy for analytic functions that obviously verify also the semi group property. So, we can write:

$$D_{\theta}^{\alpha}f(z) = \frac{\Gamma(\alpha-\beta+1)}{2\pi j}\int_{C} f^{(\beta)}(w+z)\frac{1}{w^{\alpha-\beta+1}}\,\mathrm{d}w \tag{3.61}$$

Let us choose again, $\beta = n \in \mathbf{Z}^+$ and $\varepsilon = n - \alpha > 0$. We obtain:

$$D_{\theta}^{\alpha}f(z) = \frac{\Gamma(-\varepsilon+1)}{2\pi j}\int_{C} f^{(n)}(w+z)w^{\varepsilon-1}\mathrm{d}w \tag{3.62}$$

or

$$D_{\theta}^{\alpha}f(z) = \frac{\Gamma(-\varepsilon+1)}{2\pi j}\int_{C_d} f^{(n)}(w)(w-z)^{\varepsilon-1}\mathrm{d}w \tag{3.63}$$

that can be considered as a Caputo-Cauchy derivative, provided the integral exists. The integration paths C and C_d are U-shaped lines as shown in Fig. 3.2. The representation (3.60) is valid because $f(z)$ is analytic and we assumed that the GL derivative exists. So (3.61) and (3.62) too.

Consider again the integration path in Fig. 3.3. As before, we can decompose (3.61) into three integrals along the two half-straight lines and the circle. We have, then:

$$D_{\theta}^{\alpha}f(z) = \frac{\Gamma(-\varepsilon+1)}{2\pi j}\left[\int_{C_1} + \int_{C_2} + \int_{C_3}\right] f^{(n)}(w+z)w^{\varepsilon-1}\mathrm{d}w$$

We do not need to continue because we can use (3.36). This is valid because $f(z)$ being analytic its nth order derivative is also. However it is interesting to pursue due to some interesting details. Thus we continue.

Over C_1 we have $w = x{\cdot}e^{j(\theta-\pi)}$, while over C_3 we have $w = x{\cdot}e^{j(\theta+\pi)}$, with $x \in R^+$, over C_2 we have $w = \rho e^{j\varphi}$, with $\varphi \in (\theta - \pi, \theta + \pi)$. We can write, at last:

$$\begin{aligned} D_{\theta}^{\alpha}f(z) = \frac{\Gamma(-\varepsilon+1)}{2\pi j} &= \int_{\infty}^{\rho} f^{(n)}(-x\cdot e^{j\theta}+z)e^{j\varepsilon(\theta-\pi)}x^{\varepsilon-1}\mathrm{d}x \\ &+ \frac{\Gamma(-\varepsilon+1)}{2\pi j}\int_{\infty}^{\rho} f^{(n)}(-x\cdot e^{j\theta}+z)e^{j\varepsilon(\theta+\pi)}x^{\varepsilon-1}\mathrm{d}x \\ &+ \frac{\Gamma(-\varepsilon+1)}{2\pi j} = \rho^{\varepsilon}\int_{\theta-\pi}^{\theta+\pi} f^{(n)}(\rho\cdot e^{j\varphi}+z)e^{j\varepsilon\varphi}j\,\mathrm{d}\varphi \end{aligned}$$

For the first and second terms, we have:

$$\int_{\infty}^{\rho} f^{(n)}(-x\cdot e^{j\theta}+z)e^{j\varepsilon(\theta-\pi)}.x^{\varepsilon-1}dx+\int_{\rho}^{\infty} f^{(n)}(-x\cdot e^{j\theta}+z)e^{j\varepsilon(\theta+\pi)}x^{\varepsilon-1}dx$$
$$=e^{j\varepsilon\theta}\cdot 2j\cdot\sin(\varepsilon\pi)\int_{\rho}^{\infty} f^{(n)}(-x\cdot e^{j\theta}+z)x^{\varepsilon-1}dx$$

For the third term, we begin by noting that the analyticity of the function $f(z)$ allows us to write:

$$f^{(n)}(-x\cdot e^{j\theta}+z)=\sum_{n}^{\infty}\frac{(-1)^n(-k)_n f^{(k)}(z)}{k!}(-1)^k x^{k-n}e^{jk\theta} \tag{3.64}$$

for $x < r \in \mathbf{R}^+$. We have then

$$j\rho^{\varepsilon}\int_{\theta-\pi}^{\theta+\pi} f^{(n)}(\rho\cdot e^{j\varphi}+z)e^{j\varepsilon\varphi}d\varphi=j\rho^{\varepsilon}\sum_{n}^{\infty}\frac{(-1)^n(-k)_n f^{(k)}(z)}{k!}(-1)^k\rho^{k-n}\int_{\theta-\pi}^{\theta+\pi} e^{j(k+\varepsilon)\varphi}d\varphi$$

Performing the integration, we have:

$$j\rho^{\varepsilon}\int_{\theta-\pi}^{\theta+\pi} f^{(n)}(\rho\cdot e^{j\varphi}+z)e^{j\varepsilon\varphi}d\varphi=2j\cdot e^{j\varepsilon\theta}\sin(\varepsilon\pi)$$
$$\times\sum_{n}^{\infty}\frac{(-1)^n(-k)_n f^{(k)}(z)}{k!}(-1)^k\frac{e^{jk\theta}\rho^{k-n+\varepsilon}}{(k+\varepsilon)}$$

As ρ decreases to zero, the summation in the last expression goes to zero. This means that when $\rho\to 0$

$$D_{\theta}^{\alpha}f(z)=e^{j\varepsilon\theta}\frac{1}{\Gamma(\varepsilon)}\int_{0}^{\infty} f^{(n)}(-x\cdot e^{j\theta}+z)x^{\varepsilon-1}dx \tag{3.65}$$

This can be considered as a generalised Caputo derivative. In fact, with $\theta=0$, we obtain:

$$D_{f}^{\alpha}f(z)=\frac{1}{\Gamma(\varepsilon)}\int_{0}^{\infty} f^{(n)}(z-x)\,x^{\varepsilon-1}dx=\frac{1}{\Gamma(\varepsilon)}\int_{-\infty}^{z} f^{(n)}(\tau)\,(z-\tau)^{\varepsilon-1}d\tau \tag{3.66}$$

that is the forward Caputo derivative in **R**.

Now, return to (3.61) and put $\beta=0$ and $\alpha=n-\varepsilon$, with $\varepsilon>0$, again:

$$D^{\alpha}_{\theta}f(z) = \frac{\Gamma(\alpha+1)}{2\pi j}\int\limits_{C_d} f(w)\frac{1}{(w-z)^{n-\varepsilon+1}}\mathrm{d}w$$
$$= \frac{\Gamma(\alpha+1)}{2\pi j}\int\limits_{C} f(w)(w-z)^{\varepsilon-n-1}\mathrm{d}w \tag{3.67}$$

But, as

$$(w-z)^{\varepsilon-n-1} = \frac{1}{(1-\varepsilon)_n}D^n_z(w-z)^{\varepsilon-1} \tag{3.68}$$

we obtain, by commuting the operations of derivative and integration

$$D^{\alpha}_{\theta}f(z) = D^n\left[\frac{\Gamma(-\varepsilon+1)}{2\pi j}\int\limits_{C} f(w)\,(w-z)^{\varepsilon-1}\mathrm{d}w\right] \tag{3.69}$$

We may wander about the validity of the above commutation. We remark that the resulting integrand function has a better behaviour than the original, ensuring that we gain something on doing such operation. The formula (3.69) is the complex version of the Riemann–Liouville derivative that we can write in the format

$$D^{\alpha}_{\theta}f(z) = D^n\left[\frac{\Gamma(-\varepsilon+1)}{2\pi j}\int\limits_{C} f(w+z)\,w^{\varepsilon-1}\mathrm{d}w\right] \tag{3.70}$$

Using again the Hankel integration path, we obtain easily:

$$D^{\alpha}_{\theta}f(z) = e^{j\varepsilon\theta}D^n\left[\frac{1}{\Gamma(\varepsilon)}\int\limits_0^{\infty} f(-x\cdot e^{j\theta}+z)x^{\varepsilon-1}\mathrm{d}x\right] \tag{3.71}$$

that is a generalised RL derivative. With $\theta = 0$, we can obtain the usual "left" formulation of the RL in *R*. With $\theta = \pi$, we obtain aside a factor the "right" RL derivative.

3.7.2 Half Plane Derivatives

Let us assume that $f(z) \equiv 0$ for $Re(z) < 0$. In this case, the summation in (2.32) {see (3.59) and (3.60)} goes only to $K = \lfloor Re(z)/Re(h) \rfloor$ and the integration path in (3.27) is finite, closed and completely in the right half complex plane. In Fig. 3.4 we assumed that z and h are real.

Consider a sequence h_n ($n = 1, 2, 3, \ldots$) going to zero. The number of poles inside the integration path is *K*, but in the limit, the quotient of two gamma functions will give rise to a multivalue expression that forces us to insert a branch

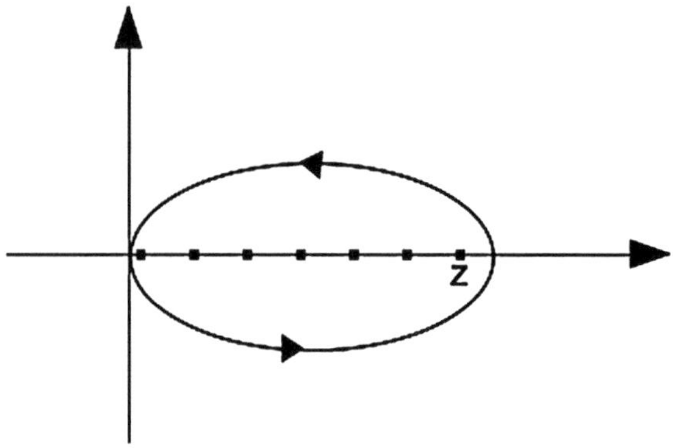

Fig. 3.4 The contour used in computing the half plane derivatives

cut line that starts at z and ends at $-\infty$. Over this line the integrand is not continuous. So, we obtain:

$$D_\theta^\alpha f(z) = \frac{\Gamma(\alpha+1)}{2\pi j}\int_C f(w)\frac{1}{(w-z)^{\alpha+1}}\mathrm{d}w + \frac{\Gamma(\alpha+1)}{2\pi j}\int_\gamma f(w)\frac{1}{(w-z)^{\alpha+1}}\mathrm{d}w \tag{3.72}$$

where C is an open contour that encircles the branch cut line and γ is a small line passing at $w = 0$ whose length we will reduce to zero {see Fig. 3.5}. However, we prefer to use the analogue to the Hankel contour. The contour γ is a short straight line over the imaginary axis. Although the integrand is not continuous, the phase has a $2\pi(\alpha+1)$ jump, the second integral above is zero. To compute the others, we are going to do a translation to obtain an integral similar to the used above.

As before and again for reducing steps, we will assume already that the straight lines are infinitely near to each other. We have, then:

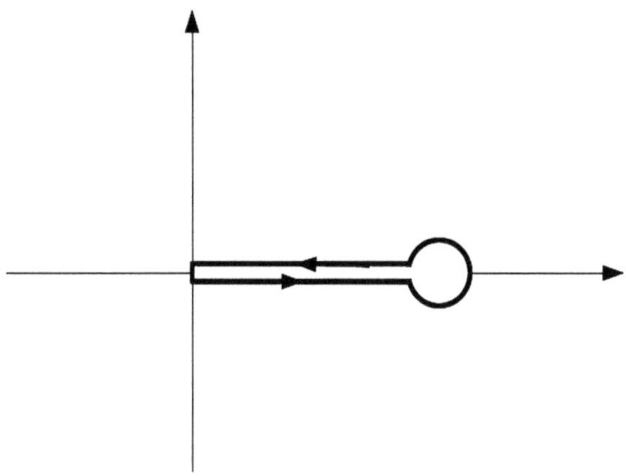

Fig. 3.5 The Hankel contour used in computing the derivative defined in Eq. 3.73

$$D_{\theta}^{\alpha}f(z) = \frac{\Gamma(\alpha+1)}{2\pi j}\left[\int_{C_1} + \int_{C_2} + \int_{C_3}\right] f(w+z)\frac{1}{w^{\alpha+1}}\mathrm{d}w \tag{3.73}$$

Over C_1 we have $w = xe^{j(\theta-\pi)}$, while over C_3 we have $w = x{\cdot}e^{j(\theta+\pi)}$, with $x \in \mathbf{R}^{+}$, over C_2 we have $w = \rho e^{j\varphi}$, with $\varphi \in (\theta - \pi, \theta + \pi)$.

Let $\zeta = |z|$. We can write, at last:

$$\begin{aligned} D_{\theta}^{\alpha}f(z) &= \frac{\Gamma(\alpha+1)}{2\pi j}\int_{\zeta}^{\rho} f(-x\cdot e^{j\theta}+z)\frac{e^{-j\alpha(\theta-\pi)}}{x^{\alpha+1}}\mathrm{d}x \\ &+ \frac{\Gamma(\alpha+1)}{2\pi j}\int_{\rho}^{\zeta} f(-x\cdot e^{j\theta}+z)\frac{e^{-j\alpha(\theta+\pi)}}{x^{\alpha+1}}\mathrm{d}x \\ &+ \frac{\Gamma(\alpha+1)}{2\pi j}\frac{1}{\rho^{\alpha}}\int_{\theta-\pi}^{\theta+1} f(\rho\cdot e^{j\varphi}+z)e^{-j\alpha\varphi}j\,\mathrm{d}\varphi \end{aligned} \tag{3.74}$$

For the first term, we have:

$$\int_{\zeta}^{\rho} f(x\cdot e^{j(\theta-\pi)}+z)\frac{e^{-j\alpha(\theta-\pi)}}{x^{\alpha+1}}\mathrm{d}x + \int_{\rho}^{\zeta} f(-x\cdot e^{j(\theta+\pi)}+z)\frac{e^{-j\alpha\theta}}{x^{\alpha+1}}\mathrm{d}x$$

$$\begin{aligned} &= e^{-j\alpha\theta}\cdot\left[e^{j\pi\alpha} - e^{-j\pi\alpha}\right]\int_{\rho}^{\zeta} f(-x\cdot e^{j\theta}+z)\frac{1}{x^{\alpha+1}}\mathrm{d}x \\ &= e^{-j\alpha\theta}\cdot 2j\cdot\sin(\alpha\pi)\int_{\rho}^{\zeta} f(-x\cdot e^{j\theta}+z)\frac{1}{x^{\alpha+1}}\mathrm{d}x \end{aligned} \tag{3.75}$$

For the second term, we have

$$j\frac{1}{\rho^{\alpha}}\int_{\theta-\pi}^{\theta+\pi} f(\rho\cdot e^{j\varphi}+z)e^{-j\alpha\varphi}\mathrm{d}\varphi = j\frac{1}{\rho^{\alpha}}\sum_{0}^{\infty}\frac{f^{(n)}(z)}{n!}\rho^{n}\int_{\theta-\pi}^{\theta+\pi} e^{j(n-\alpha)\varphi}\mathrm{d}\varphi \tag{3.76}$$

Performing the integration, we have:

$$j\frac{1}{\rho^{\alpha}}\int_{\theta-\pi}^{\theta+\pi} f(\rho\cdot e^{j\varphi}+z)e^{-j\alpha\varphi}\mathrm{d}\varphi = -2j\cdot e^{-j\alpha\theta}\sin(\alpha\pi)\sum_{0}^{\infty}\frac{f^{(n)}(z)}{n!}\frac{e^{j(n-\alpha)\theta}\rho^{n-\alpha}}{(n-\alpha)} \tag{3.77}$$

As before:

$$\sum_{0}^{\infty}\frac{f^{(n)}(z)}{n!}\cdot\frac{e^{j(n-\alpha)\theta}\rho^{n-\alpha}}{(n-\alpha)}=\left[-\sum_{0}^{\infty}\frac{f^{(n)}(z)}{n!}e^{jn\theta}\int_{\rho}^{\infty}x^{n-\alpha-1}dx+\sum_{N+1}^{\infty}\frac{f^{(n)}(z)}{n!}\frac{e^{jn\theta}\rho^{n-\alpha}}{(n-\alpha)}\right]$$

Substituting it in (3.77) and joining to (3.75) we can write:

$$D_{\theta}^{\alpha}f(z)=K\int_{\rho}^{\zeta}\frac{\left[f(-x\cdot e^{j\theta}+z)-\sum_{0}^{N}\frac{f^{(n)}(z)}{n!}e^{jn\theta}x^{n}\right]}{x^{\alpha+1}}dx$$
$$-K\cdot\sum_{N+1}^{\infty}\frac{f^{(n)}(z)}{n!}(-1)^{n}\frac{\rho^{n-\alpha}}{(n-\alpha)}+\Theta \tag{3.78}$$

with

$$\Theta=-\sum_{0}^{N}\frac{f^{(n)}(z)}{n!}e^{jn\theta}\int_{\zeta}^{\infty}x^{n-\alpha-1}dx=z^{-\alpha}\sum_{0}^{N}\frac{f^{(n)}(z)}{n!}\frac{z^{n}}{n-\alpha}$$

If $\alpha<0$, we make the three summations equal to zero. Using the reflection formula of the gamma function we obtain for K

$$K=-\frac{\Gamma(\alpha+1)e^{-j\theta\alpha}}{\pi}\sin(\alpha\pi)=\frac{e^{-j\theta\alpha}}{\Gamma(-\alpha)} \tag{3.79}$$

Now let ρ go to zero. The second term on the right hand side in (3.78) goes to zero and we obtain:

$$D_{\theta}^{\alpha}f(z)=K\cdot\int_{0}^{\zeta}\frac{\left[f(-x\cdot e^{j\theta}+z)-\sum_{0}^{N}\frac{f^{(n)}(z)}{n!}e^{jn\theta}x^{n}\right]}{x^{\alpha+1}}dx+\sum_{0}^{N}\frac{f^{(n)}(z)}{n!}\frac{z^{n-\alpha}}{n-\alpha} \tag{3.80}$$

This result shows that in this situation and with $\alpha>0$ we have a regularised integral and an additional term. This means that it is somehow difficult to compute the fractional derivative by using (3.80): a simple expression obtained from the general GL derivative

$$D_{\theta}^{\alpha}f(z)=e^{-j\theta\alpha}\lim_{|h|\to 0}\frac{\sum_{k=0}^{\lfloor\zeta/h\rfloor}(-1)^{k}\binom{\alpha}{k}f(z-kh)}{|h|^{\alpha}} \tag{3.81}$$

leads to a somehow complicated formation in (3.80). However, if $\alpha<0$ we obtain:

$$D_{\theta}^{\alpha}f(z)=K\cdot\int_{0}^{\zeta}\frac{f(x\cdot e^{j\theta}+z)}{x^{\alpha+1}}dx \tag{3.82}$$

So, we must avoid (3.80). To do it, remark first that, from (3.58) we have:

$$D_{\theta f}^{\alpha}(z) = D_{\theta}^{n}\left[D_{\theta}^{-\varepsilon}f(t)\right] = D_{\theta}^{-\varepsilon}\left[D_{\theta f}^{n}(t)\right] \tag{3.83}$$

This means that we can compute the α order derivative into two steps. As one step is a fractional anti-derivative, we avoid (3.80) and use (3.82). The order of the steps: computing the integer order derivative before or after the anti-derivative leads to

$$D_{\theta f}^{\alpha}(z) = K \cdot \int_{0}^{\zeta} \frac{f^{(n)}(x \cdot e^{j\theta} + z)}{x^{\alpha+1}} \mathrm{d}x \tag{3.84}$$

and

$$D_{\theta f}^{\alpha}(z) = K \cdot D^{n} \int_{0}^{\zeta} \frac{f(x \cdot e^{j\theta} + z)}{x^{\alpha+1}} \mathrm{d}x \tag{3.85}$$

that are the C and RL formulations in the complex plane. However, from (3.83) we can write also:

$$D_{\theta f}^{\alpha}(z) = e^{-j\theta\alpha} \lim_{|h|\to 0} \frac{\sum_{k=0}^{\lfloor \zeta/h \rfloor} (-1)^{k} \binom{-\varepsilon}{k} f^{(n)}(z - kh)}{|h|^{\alpha}} \tag{3.86}$$

and

$$D_{\theta f}^{\alpha}(z) = \left[e^{-j\theta\alpha} \lim_{|h|\to 0} \frac{\sum_{k=0}^{\lfloor \zeta/h \rfloor} (-1)^{k} \binom{-\varepsilon}{k} f(z - kh)}{|h|^{\alpha}} \right]^{(n)} \tag{3.87}$$

These results mean that: we can easily define C-GL (3.86) and RL-GL (3.87) derivatives. Attending to the way we followed for going from GL to C and RL, we can conclude that, in the case of analytic functions, the existence of RL or C derivatives ensure the existence of the corresponding GL. The reverse may be not correct, since the commutation of limit and integration in (3.22) and (3.23) may not be valid. It is a simple task to obtain other decompositions of α, leading to valid definitions. One possibility is the Miller-Ross sequential differintegration [7]:

$$D^{\alpha}x(t) = \left[\prod_{i=1}^{N} D^{\sigma_i}\right] x(t) \tag{3.88}$$

with $\alpha = N\sigma$. This is a special case of multi-step case proposed by Samko et al. [5] and based on the Riemann–Liouville definition:

$$D^{\alpha}x(t) = \left[\prod_{i=1}^{N} D^{\sigma_i}\right] x(t) \tag{3.89}$$

with

$$\alpha = \left[\sum_{i=1}^{N} \sigma_i\right] - 1 \quad \text{and} \quad 0 < \sigma_i \leq 1 \tag{3.90}$$

These definitions suggest us that, to compute a α derivative, we have infinite ways, depending on the steps that we follow to go from 0 (or $-v$) to α, that is we express α as a summation of N reals σ_i ($i = 0, \ldots, N - 1$), with the σ_i not necessarily less or equal to one.

3.8 Conclusions

We started from the Grünwald–Letnikov derivative, obtained an integral formulation representing the summation. From it and by permuting the limit and the integration we obtained the general Cauchy derivative. From this and using the Hankel contour as integration path we arrived at a regularized derivative. This is similar to Hadamard definition but was obtained without rejecting and infinite part. With the presented methodology we could also obtain the Riemann–Liouville and Caputo derivatives and showed also that they can be computed with the Grünwald–Letnikov definition.

References

1. Diaz JB, Osler TJ (1974) Differences of fractional order. Math Comput 28(125):185–202
2. Kiryakova V (1996) A long standing conjecture failed? In: Proceedings of the 2nd international workshop on transform methods & special functions, Varna'96 Aug 1996
3. Ortigueira MD (2004) From differences to differintegrations. Fract Calc Appl Anal 7(4):459–471
4. Ortigueira MD (2006) A coherent approach to non integer order derivatives. Signal Process, Special Section Fractional Calculus Applications in Signals and Systems. 86(10):2505–2515
5. Samko SG, Kilbas AA, Marichev OI (1987) Fractional integrals and derivatives—theory and applications. Gordon and Breach Science Publishers, New York
6. Nishimoto K (1989) Fractional calculus. Descartes Press Co., Koriyama
7. Miller KS, Ross B (1993) An introduction to the fractional calculus and fractional differential equations. Wiley, New York
8. Podlubny I (1999) Fractional differential equations: an introduction to fractional derivatives, fractional differential equations, to methods of their solution and some of their applications. Academic Press, San Diego
9. Hoskins RF, Pinto JS (1994) Distributions ultradistributions and other generalised functions. Ellis Horwood Limited, Chichester

10. Zemanian AH (1987) Distribution theory and transform analysis. Dover Publications, New York
11. Poularikas AD (ed) (2000) The transforms and applications handbook. CRC Press, Boca Raton
12. Chaudhry MA, Zubair SM (2002) On a class of incomplete gamma functions with applications. Chapman Hall, London
13. Henrici P (1991) Applied and computational complex analysis, vol 2. Wiley, New York, pp 389–391
14. Ortigueira MD (2005) Two new integral formulae for the Beta function. Int J Appl Math 18(1):109–116
15. Kilbas AA, Srivastava HM, Trujillo JJ (2006) Theory and applications of fractional differential equations. Elsevier, Amsterdam

Chapter 4
Fractional Linear Shift-Invariant Systems

4.1 Introduction

The applications of Fractional Calculus to physics and engineering are not recent: the beginning of the application to viscosity dates back to the thirties in the past century. During the last 20 years the application domains of fractional calculus increased significantly: seismic analysis [1], dynamics of motor and premotor neurones of the oculomotor systems [2], viscous damping [3, 4], electric fractal networks [5], fractional order sinusoidal oscillators [6] and, more recently, control [7, 8, 9], and robotics [10]. One of the areas where such can be verified is the Biomedical Engineering [11, 12]. The now classic fractional Brownian motion (fBm) modeling is an application of the fractional calculus [13–15]. We define a fractional noise that is obtained through a fractional derivative of white noise. The fBm is an integral of the fractional noise.

Although the fractional linear systems have an already long history, the first formal presentation of the Fractional Linear System Theory was done in 2000 [16]. Most of elementary books on Signals and Systems consider only the integer derivative order case and treat the corresponding systems, studying their impulse, step and frequency responses and their Transfer Function. It is not such as simple matter, if one substitutes fractional derivatives for the common derivatives. The objective of this chapter is to treat the Fractional Continuous-Time Linear Shif-Invariant Systems as it is done with the usual systems. As we will see the approach deals with very well known concepts. We merely generalise them to the fractional case.

M. D. Ortigueira, *Fractional Calculus for Scientists and Engineers*,
Lecture Notes in Electrical Engineering, 84, DOI: 10.1007/978-94-007-0747-4_4,

4.2 Description

The most common and useful continuous-time linear systems are the lumped parameter systems that are described by linear differential equations. The simplest of these systems are the integrators, differenciators and constant multipliers (amplifiers/attenuators). The referred lumped parameter linear systems are associations (cascade, parallel or feedback) of those simple systems. Here, we will study the systems that result from the use of fractional differintegrators or integrators and that are described by linear fractional differential equations. For now, we will assume that the coefficients of the equation are constant, so the corresponding system will be a fractional linear time-invariant (FLTI) system. With this definition, we are in conditions to define and compute the *Impulse Response and Transfer Function.*

According to what we just said, we will consider FLTI systems described by a differential equation with the general format:

$$\sum_{n=0}^{N} a_n D^{\nu_n} y(t) = \sum_{m=0}^{M} b_m D^{\nu_m} x(t) \tag{4.1}$$

where the ν_n are the differintegration orders that, in the general case, are complex numbers. Here, we will assume they are positive real numbers. Let $h(t)$ be the output of the system to the impulse (impulse response)

$$\sum_{n=o}^{N} a_n D^{\nu_n} h(t) = \sum_{m=0}^{M} b_m D^{\nu_m} \delta(t) \tag{4.2}$$

and convolve both members in (4.2) with $x(t)$

$$\sum_{n=0}^{N} a_n D^{\nu_n} h(t) * x(t) = \sum_{m=0}^{M} b_m D^{\nu_m} \delta(t) * x(t)$$

As known, $x(t) * \delta(t) = x(t)$ for almost all the interesting functions, namely for tempered distributions that we will assume to deal with. On the other hand

$$[D^{\alpha} h(t)] * x(t) = D^{\alpha}[h(t) * x(t)]$$

and comparing with (4.1) we conclude that the output is given by $y(t) = x(t) * h(t)$, a well known result. This brings an important consequence:

*The exponentials defined in **R** are eigenfunctions of the **FLTI** systems.*
Let us see why. Assume that $x(t) = e^{st}$ with $t \in R$. Then

$$y(t) = e^{st} \int_{-\infty}^{+\infty} h(\tau) e^{-s\tau}\, d\tau = H(s) \cdot e^{st} \tag{4.3}$$

where $H(s)$ is the transfer function that is the LT of the impulse response as in the integer order case. Inserting $x(t) = e^{st}$ and $y(t)$ as given by (4.3) into (4.2) we obtain an explicit representation for the transfer function

$$H(s) = \frac{\sum_{m=0}^{M} b_m s^{\nu_m}}{\sum_{n=0}^{N} a_n s^{\nu_n}} \tag{4.4}$$

provided that $\mathrm{Re}(s) > 0$ or $\mathrm{Re}(s) < 0$. So (4.4) represents two different systems depending on the adopted region of convergence. In the following we will assume to deal with causal systems, $\mathrm{Re}(s) > 0$.

If we let $s \to j\omega$ and use the results obtained in Sect. 2.7.4 we obtain the Frequency Response, $H(j\omega)$,

$$H(s) = \frac{\sum_{m=0}^{M} b_m (j\omega)^{\nu_m}}{\sum_{n=0}^{N} a_n (j\omega)^{\nu_n}} \tag{4.5}$$

We must be careful about the definition of the fractional power of $j\omega$ (see Sect. 2.7.4). With (4.5) we can get the Bode diagrams as in the usual systems. It is interesting to remark that the asymptotic amplitude Bode diagrams are constituted by straight lines with slopes that, at least in principle, may assume any value, contrarily to the usual case where the slopes are multiples of 20 dB/decade.

4.3 From the Transfer Function to the Impulse Response

To obtain the Impulse Response from the Transfer Function we proceed almost as usually. However, we must be careful. Let us begin by considering the simple case of a differintegrator:

$$H(s) = s^{\alpha}, \quad \alpha \neq 0 \tag{4.6}$$

s^{α} is a multivalued expression defining an infinite number of Riemann surfaces. Each Riemann surface defines one function. Therefore, (4.6) can represent an infinite number of linear systems. However, only the principal Riemann surface, $\{z : -\pi < \arg(z) \leq \pi\}$, may lead to a real system. Constraining this function by imposing a region of convergence, we define a transfer function. The impulse response was computed in Chap. 2 and is given by:

$$D_f^{\alpha}\delta(t) = \frac{t^{-\alpha-1}}{\Gamma(-\alpha)} u(t) \tag{4.7}$$

Now return to (4.4). The general case is hard to solve because it is difficult to find the poles of the system. For now, we will consider the simpler case where

(a) the ν_n are rational numbers that we will write in the form p_n/q_n. Let p and q be the least common multiples of the p_n and q_n, then $\nu_n = np/q$, where n and

q are positive integer numbers. So, $\nu_n = n \cdot \nu$, with $\nu = 1/q$ (a differential equation with $\nu = 1/2$ is said semi-differential). The coefficients and orders do not coincide necessarily with the previous ones, since some of the coefficients can be zero[1]

(b) the ν_n are irrational numbers but multiples of a $\nu, 0 \le \nu \le 1$. When comparing with the integer order case, we performed a substitution $s \to s^\nu$. This implies that the interval $[0, \pi)$ is transformed into the interval $[0, \pi\nu)$, meaning that $\nu \le 1$

Then, Eqs. 4.1 and 4.4 assume the forms:

$$\sum_{n=0}^{N} a_n D^{n\nu} y(t) = \sum_{m=0}^{M} b_m D^{m\nu} x(t) \tag{4.8}$$

and

$$H(s) = \frac{\sum_{m=0}^{M} b_m s^{m\nu}}{\sum_{n=0}^{N} a_n s^{n\nu}} \tag{4.9}$$

With a Transfer Function as in (4.9) we can perform the inversion quite easily, by following the steps:

Transform $H(s)$ into $H(z)$, by substitution of s^ν for z[2]

(1) The denominator polynomial in $H(z)$ is the indicial polynomial or characteristic pseudo-polynomial. Perform the expansion of $H(z)$ in partial fractions.
(2) Substitute back s^ν for z, to obtain the partial fractions in the form:

$$F(s) = \frac{1}{(s^\nu - p)^k} \quad k = 1, 2, \ldots \tag{4.10}$$

(3) Invert each partial fraction.
(4) Add the different partial Impulse Responses.

Now, go a step ahead and consider the simple case corresponding to $k = 1$ in the fraction (4.10). Let $f(t)$ be the inverse Laplace transform of $F(s) = \frac{1}{s^\nu - p}$. The $k > 1$ case in (4.10) does not present great difficulties except some additional work. We can use the convolution to solve the problem. Alternatively we can differentiate. For example:

$$\frac{1}{(s^\nu - p)^2} = -\frac{1}{\nu s}1 - \nu \frac{\mathrm{d}}{\mathrm{d}s}\left[\frac{1}{s^\nu - p}\right] \tag{4.11}$$

---

[1] For example, the equation: $[aD^{1/3} + bD^{1/2}]y(t) = x(t)$ transforms into: $[bD^{3.1/6} + aD^{2.1/6} + 0.D^{1/6}]y(t) = x(t)$.

[2] We are assuming that $H(z)$ is a proper fraction; otherwise, we have to decompose it in a sum of a polynomial (inverted separately) and a proper fraction.

$$\mathrm{LT}^{-1}\left[\frac{1}{(s^{\nu}-p)^{2}}\right]=\frac{1}{\nu}D^{1-\nu}[t^{\nu}\cdot f(t)] \tag{4.12}$$

We do not go further, since this example shows how we can proceed in the general case.

There is another alternative, possibly simpler. The inverse of $F(s)$, $f(t)$, is a function of p. To enhance this fact, write

$$f_p(t)=LT^{-1}\left[\frac{1}{(s^{\nu}-p)}\right]$$

Continue by computing the first order derivative relatively to p (denoted by D_p)

$$D_p f_p(t)=LT^{-1}\left[\frac{1}{(s^{\nu}-p)^{2}}\right]$$

This can be repeated and generalised.

Return to the $k = 1$ case and consider the denominator. The equation $s^{\nu} = p$ has infinite solutions that are on a circle of radius $|p|^{1/\nu}$. However, in the general case, we cannot assure the existence of one pole in the principal Riemann surface. If θ_0 is the argument of $\mathbf{p}$ in (4.10), we must have $|\theta_0| < \pi\nu$. This has implications in the inverse transform.

4.4 Partial Fraction Inversion

4.4.1 By the Inversion Integral

We are going to present the steps to inverting a transfer function of a fractional causal system of the type (4.9),

$$h(t)=\frac{1}{2\pi j}\int_{a-j\infty}^{a+j\infty} F(s^{\alpha})\mathrm{e}^{st}\,\mathrm{d}s \tag{4.13}$$

where we enhanced the s^{α} dependence of the function. We will assume that

$$\lim_{s\to 0} sF(s^{\alpha})=0, \quad \alpha>0 \tag{4.14}$$

and

$$\lim_{s\to\infty} F(s^{\alpha})=0, \quad \alpha>0 \tag{4.15}$$

Fig. 4.1 Integration path for inverting $F(s^{\alpha})$

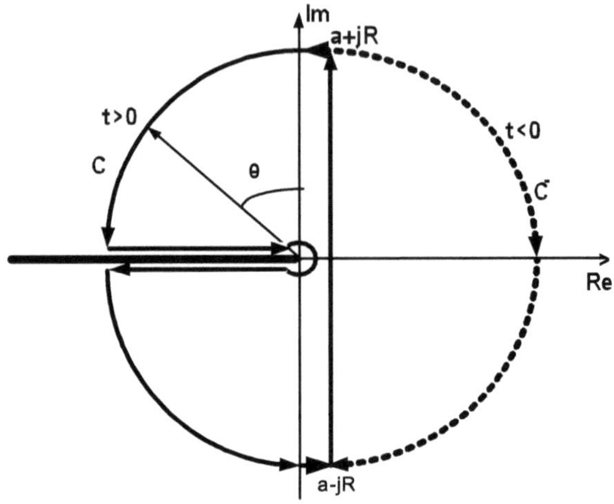

in the cut plane $|\arg(s)| < \pi$ and also that the function has N poles in the left half complex plane and in the region surrounded by the integration contour. To do the integration we used the residue theorem and the closed integration path shown in Fig. 4.1. According to the condition (4.14), the term corresponding to the short circle goes to zero as $\rho \to 0$ and the integral along the larger quarter circles go also to zero as $R \to \infty$. Reducing steps, we arrive at

$$h(t) = \sum_1^N R_i + \frac{1}{2\pi j}\int_{\infty}^{0} F\left(\sigma^{\alpha}\mathrm{e}^{-j\alpha\pi}\right)\mathrm{e}^{-\sigma t}\,\mathrm{d}\sigma + \frac{1}{2\pi j}\int_{0}^{\infty} F\left(\sigma^{\alpha}\mathrm{e}^{j\alpha\pi}\right)\mathrm{e}^{-\sigma t}\,\mathrm{d}\sigma$$

where R_i are the residues of $F(s^{\alpha})\mathrm{e}^{st}$ and finally

$$h(t) = \sum_1^N R_i - \frac{1}{2\pi j}\int_{0}^{\infty} \left[F\left(\sigma^{\alpha}\mathrm{e}^{j\alpha\pi}\right) - F\left(\sigma^{\alpha}\mathrm{e}^{-j\alpha\pi}\right)\right]\mathrm{e}^{-\sigma\tau}\,\mathrm{d}\sigma \tag{4.16}$$

This formula can be used to invert $s^{-\alpha}(\alpha > 0)$ very easily [17].

Let us apply the above formula to invert the partial fraction $\frac{1}{(s^{\alpha}-p)}$. Start by stating that we will work in the principal branch of the multivalue expression s^{α}. We have two situations: if $|\arg(p)| \le \pi\alpha$, we have a pole in the assumed domain; if $\pi\alpha < |\arg(p)| < \pi$, the function does not have any pole. Applying the above result for the first case we obtain immediately:

$$h(t) = \frac{p^{1/\alpha-1}}{\alpha}\mathrm{e}^{p^{1/\alpha}t}u(t) + \frac{1}{\pi}\int_{0}^{\infty} \frac{\sigma^{\alpha}\sin(\pi\alpha)}{\sigma^{2\alpha} - 2\sigma^{\alpha}p\cos(\pi\alpha) + p^2}\mathrm{e}^{-\sigma t}\,\mathrm{d}\sigma u(t) \tag{4.17}$$

If $\frac{1}{(s^{\alpha}-p)}$ does not have any pole, the first term on the right hand side in (4.17) is zero. The expression (4.17) is useful for numerical implementations. On the other hand it gives us a very interesting information. The impulse response corresponding to $\frac{1}{(s^\alpha - p)}$ has two terms with completely different behaviour:

(a) the first expresses the short range memory
(b) the second is concerned with the long range memory

This means that we can combine all the terms belonging to all the partial fractions of $H(s)$ into a sum of two partial transfer functions: one is a rational function of s and states the short time behaviour while the other is more involved and is responsible for the long range memory of the system.

To look for the possible pole of the partial fraction we must solve the equation

$$s^\alpha = p$$

To do it, write it in the format

$$|s|^\alpha e^{j\alpha\varphi} = |p| e^{j\arg(p)}$$

As φ must be inside the interval $(-\pi, \pi)$, $\arg(p)$ must be in the interval $(-\alpha\pi, \alpha\pi)$, since we considered the first branch of the power function. This means that we must constrain α to be in the interval (0,1].

4.4.2 By Series Expansion

To invert (4.10) when α is any real or complex it is usually used the Mittag–Leffler function. However, the so-called α-exponential function, closely related to that, is more useful [18]. We are going to see how we can obtain it. Let us return back to (4.10) with $k = 1$. As it is well known the integration path for the inverse Laplace transform is any straight line in the region of convergence. So, consider that $\mathrm{Re}(s) > |p|^{1/\alpha}$. So, in this region, we can write:

$$F(s) = s^{-\alpha} \sum_{n=0}^{\infty} p^n s^{-n\alpha} = \sum_{n=1}^{\infty} p^{n-1} s^{-n\alpha}$$

Before going further we must remark that the above expansion is valid for both causal and anti-causal cases, because it converges for $|s| > |p|^{1/\alpha}$. So we have to decide which one to use, but this also means that we can obtain the unstable solution. With this in mind, we are going to do the inversion, considering the causal solution. All the terms in the series are analytic for $\mathrm{Re}(s) > 0$ and their inverse LT is known from Chap. 2. Inverting term by term the above series we obtain the α-exponential function

$$f(t) = e_\nu(pt) = \sum_{n=1}^{\infty} \frac{p^{n-1} t^{n\alpha - 1}}{\Gamma(n\alpha)} u(t) \tag{4.18}$$

that is a generalization of the causal exponential. Expression (4.18) suggests us to work with the step response instead of the impulse response to avoid working with non-regular functions near the origin.

In deducing (4.18) we did not impose any constraint on α: it is valid for any real or complex value. However, for $\alpha \geq 2$, it does not have any interest because in this case the partial fraction (4.10) will have at least one pole in the right half plane and so the system is unstable. We will return to this subject later.

4.4.3 Rational Case

If α is a rational number of the type m/n with $m, n \in Z^+$ we can obtain a simpler expression. We are going to proceed to the inversion of the partial fraction by using a well known result referring the sum of the first n terms of a geometric sequence we obtain[3]:

$$F(s) = \frac{1}{(s^{m/n} - p)} = \frac{\sum_{k=1}^{n} p^{k-1} s^{p(1-k/n)}}{s^m - p^n} \tag{4.19}$$

As in this case we can always decompose the above fractions into a sum of m simpler fractions, we are going to consider only the $1/n$ case. We have

$$F(s) = \frac{1}{(s^{1/n} - p)} = \frac{\sum_{k=1}^{n} p^{k-1} s^{1-k/n}}{s - p^n} \tag{4.20}$$

That can be written as

$$F(s) = \frac{1}{(s^{1/n} - p)} = \frac{p^{n-1}}{s - p^n} + \frac{\sum_{k=1}^{n-1} p^{k-1} s^{1-k/n}}{s - p^n} \tag{4.21}$$

We conclude that the LT inverse of a partial fraction as $F(s) = \frac{1}{s^{1/n} - p}$ is the sum of an exponential $p^{q-1} e^{p^n \cdot t} \cdot u(t)$ and a linear combination of its $n-1$ fractional derivatives of orders $1 - k/n$, $k = 1, 2, \ldots, n-1$. These can be computed using the rules for the derivatives of the causal power function presented in Chap. 1. This result agrees with (4.17).

Caution: (4.21) must only be used when "p" is a pole, otherwise, we are using derivatives of an increasing exponential function. Although we expect that the resulting function goes to zero with increasing t, this may not happen for numerical reasons. Even a direct implementation of (4.18) has severe convergence problems.

---

[3] With reason $r = b/x$, we obtain: $\sum_{j=0}^{n-1} r^j = \frac{1-r^n}{1-r} \Rightarrow \sum_{j=0}^{n-1} b^j \cdot x^{-j} = \frac{1-b^n . x^{-n}}{1-b/x}$ or $x^n - b^n = (x-b) \cdot \sum_{j=1}^{n} b^{j-1} \cdot x^{n-j}$ from where $\frac{1}{x-b} = \frac{\sum_{j=1}^{n} b^{j-1} \cdot x^{n-j}}{x^n - b^n}$.

4.5 Stability of Fractional Linear Time Invariant Continuous-Time Systems

The study of the stability of the FLTI systems we are going to do is based on the BIBO[4] stability criterion that implies stability when the impulse response is absolutely integrable.

The simplest FLTI system is the system with transfer function $H(s) = s^{\nu}$ with s belonging to the principal Riemann surface. If $\nu > 0$, the system is definitely unstable, since the impulse response is not absolutely integrable, even in a finite interval. If $-1 < \nu < 0$, the impulse response remains a limited function when t increases indefinitely and it is absolutely integrable in every finite interval. Therefore, we will say that the system is wide sense stable. This case is interesting to the study of the fractional stochastic processes. If $\nu = -1$, the normal integrator, the system is wide sense stable. The case $\nu < -1$ corresponds to an unstable system, since the impulse response is not a limited function when t goes to $+\infty$.

Consider the LTI systems with transfer function $H(s)$ a quotient of two polynomials in s^{ν}. The transformation $w = z^n$, transforms the sector $0 \le \theta \le 2\pi/n\{\theta = \arg(z)\}$ into the entire complex plane. So, the sector $\frac{\pi}{2n} \le \theta \le \frac{\pi}{2n} + \frac{\pi}{n}$ is transformed in the left half plane. Consider the first Riemann surface of $z = s^{\nu}$ defined by $\theta = \arg(s) \in (-\pi, \pi]$. This domain is transformed into $\varphi = \arg(z) \in (-\pi\alpha, \pi\alpha]$. However the poles leading to instability must be inside the sector $(-\pi\alpha/2, \pi\alpha/2)$. For each p in (4.10) we have two situations leading to stability:

- There is no pole inside the sector $(-\pi, \pi]$. This happens when $\arg(p) > \pi\alpha$.
- There is a pole in one of the sectors:$(-\pi\alpha, -\pi\alpha/2)$ and $(\pi\alpha/2, \pi\alpha)$.

The poles with argument equal to $\pm\pi\alpha/2$ may lead to wide sense stable systems as in the usual systems. These conclusions come from the development we did in Sect. 4.4 and lead us to conclude that we must have $0 < \alpha \le 1$. However we can enlarge the interval of stability. Consider the transfer function $H(s) = \frac{1}{s^{\alpha}-p}$, with $1 < \alpha < 2$. As $(s^a - p) = \left(s^{a/2} - p^{1/2}\right)\left(s^{a/2} + p^{1/2}\right)$. This means that $\frac{1}{s^{\alpha}-p}$ can be decomposed into two partial fractions, both stable, although only one of them has a pole in its first Riemann sheet; the other has no pole inside. This means that $\frac{1}{s^{\alpha}-p}$ can be stable for $1 < \alpha < 2$, but we must care when doing the LT inversion. Any way, we avoid unwanted troubles by decomposing the denominator of the transfer function until all the partial fractions of the type $\frac{1}{s^{\alpha}-p}$ have $\alpha \le 1$.

We are going to exemplify the situation where a system goes from stability to instability with increasing order.

The "single-degree-of-freedom fractional oscillator" consists of a mass and a fractional Kelvin element and it is applied in viscoelasticity. The equation of motion is

[4] Bounded input, bounded output.

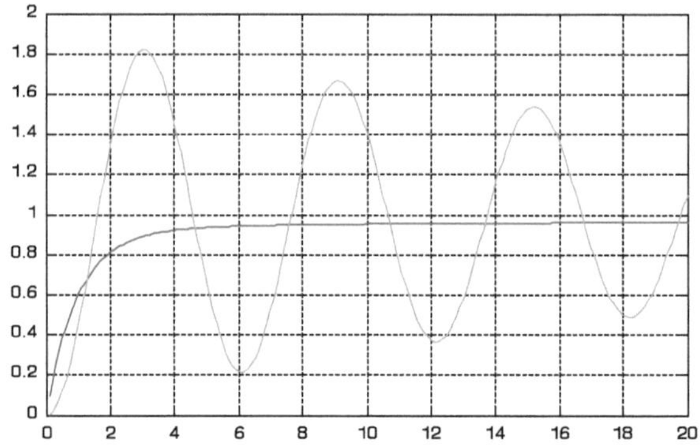

Fig. 4.2 Step responses of system defined by (4.23) for $\alpha = 1/4$ and 1/2

$$mD^2x(t) + cD^\alpha x(t) + kx(t) = f(t) \tag{4.22}$$

where m is the mass, c the damping constant, k the stiffness, x the displacement and f the forcing function. We are going to do it. Let us introduce the parameters: $\omega_0 = \sqrt{k/m}$ as the undamped natural frequency of the system and $\zeta = \frac{c}{2m\omega_0^{2-\alpha}}$. We rewrite the above equation in the form:

$$D^2x(t) + 2\omega_0^{2-\alpha}\zeta D^\alpha x(t) + \omega_0^2 x(t) = f(t) \tag{4.23}$$

The transfer function is

$$H(s) = \frac{1}{s^2 + 2\omega_0^{2-\alpha}\zeta s^\alpha + \omega_0^2} \tag{4.24}$$

with indicial polynomial $s^4 + 2\omega_0^{3/2}\zeta s + \omega_0^2$. Its roots can be found by a standard procedure, but it is a bit difficult to get useful conclusions. However, as the coefficients in s^3 and s^2 are zero, we can conclude that four roots are on two vertical straight lines with symmetric abscissas. For example, with $\omega_0 = 1$ rad/s and $\zeta = 0.05$ the roots are: $s_1 = 0.7073 + j0.7319, s_2 = 0.7073 - j0.7319, s_3 = -0.7073 + j0.6819$, and $s_4 = -0.7073 - 0.6819$, and have arguments: ± 45.9776 and ± 136.0492. This means that for α such that $\alpha\pi/2 > 45.9776/180$ we have instability. In Fig. 4.2 we present the step responses for $\alpha = 1/4$ and $\alpha = 1/2$. The oscillating response corresponds to $\alpha = 1/2$.

4.6 Examples of Simple FLTI Systems

We are going to present two simple examples obtained by adding and subtracting two partial fractions corresponding to conjugate poles:

$$H_1(s) = \frac{2s^\alpha - p - p^*}{s^{2\alpha} - (p + p^*)s^\alpha + |p|^2} \quad \text{and} \quad H_2(s) = \frac{p - p^*}{s^{2\alpha} - (p + p^*)s^\alpha + |p|^2} \tag{4.25}$$

In Fig. 4.3 we present the step responses corresponding to orders $\alpha_n = n/10$, with $n = 1, 2, \ldots, 10$.

The corresponding Bode plots are shown in Fig. 4.4. The upper plots refer to the first system while the lower correspond to the second one. We see that the plots become flatter with decreasing orders.

It is interesting to study typical electric circuits and get a comparison with well known responses. Begin with lowpass RC circuit with a transfer function that is a simple fraction:

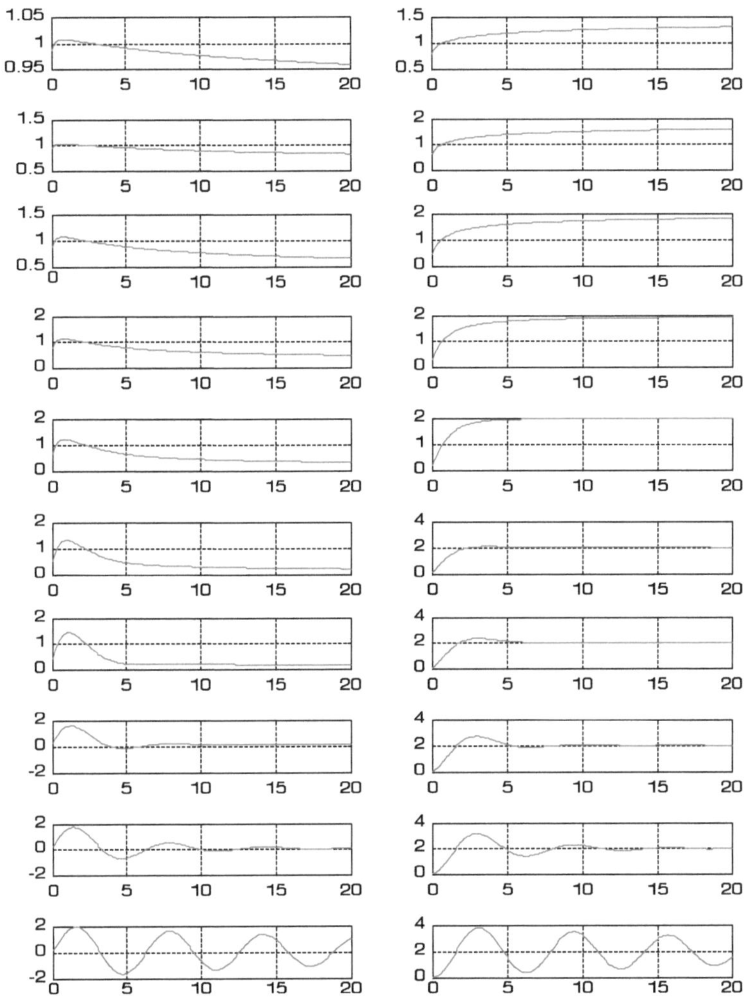

Fig. 4.3 Step responses of systems defined in (4.25) for $\alpha_n = n/10$, $n = 1,2,\ldots,10$. The *left* column represents responses of the system defined by $H_1(s)$ and the *right* one those of $H_2(s)$

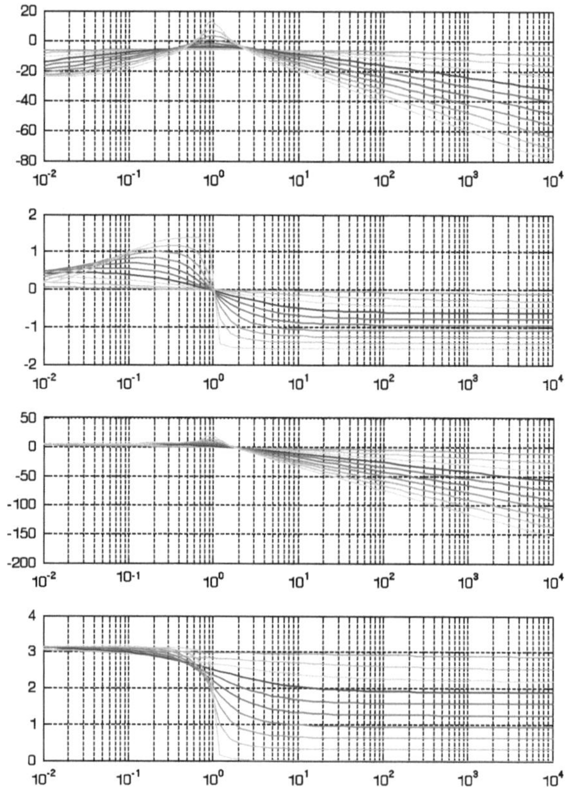

Fig. 4.4 Bode plots of systems defined in (4.25) for $\alpha_n = n/10$, $n = 1,2,\ldots,10$, The *upper plots* refer to the system defined by $H_1(s)$ and the and the *lower* to $H_2(s)$

$$H(s) = \frac{1}{s^\alpha + 1}, \quad 0 < \alpha < 1 \tag{4.26}$$

In Figs. 4.5, 4.6, 4.7, 4.8, we can see the frequency and time responses of the system for $\alpha = 0.5, \ldots, 1$. In the figures 4.9, 4.10, 4.11, 4.12, we depict the results of similar study of the L series R//C circuit (C fractional).

4.7 Initial Conditions

4.7.1 Introduction

The initial value problem is a theme that remains quite up-to-date, even in the classic integer order case [19]. In fact, the computation of the output of a linear system under a given set of initial conditions is an important task in daily applications. Traditionally this task has been accomplished by means of the unilateral

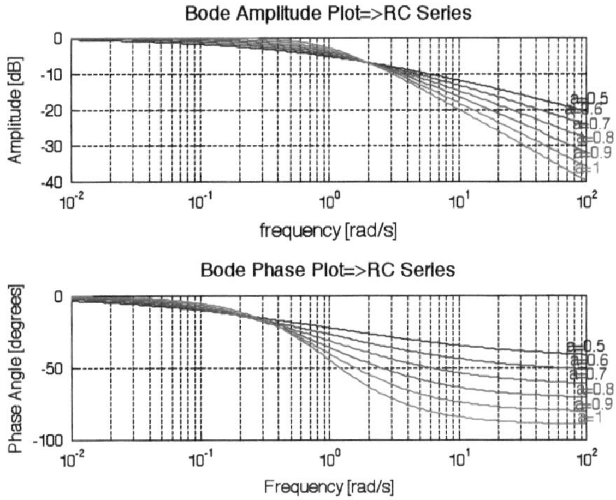

Fig. 4.5 Bode plots for the fractional RC circuit for $\alpha = 0.5, \ldots, 1$

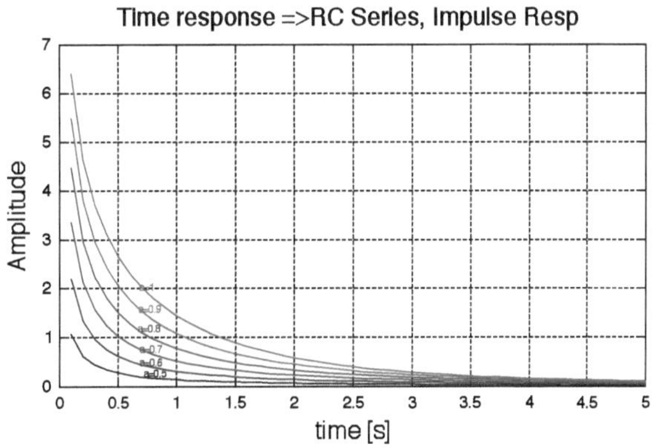

Fig. 4.6 Impulse responses of the fractional RC circuit for $\alpha = 0.5, \ldots, 1$

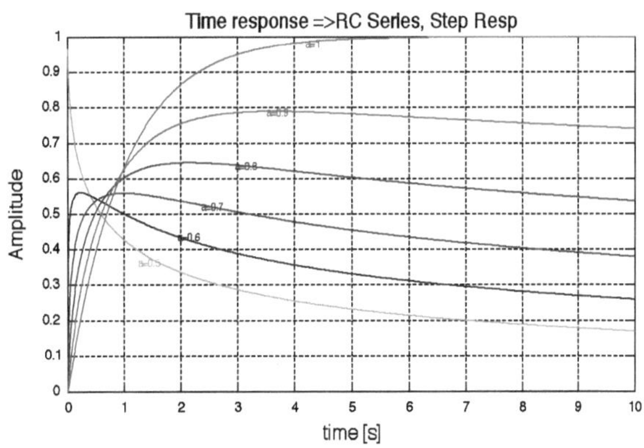

Fig. 4.7 Step responses of the fractional RC circuit for $\alpha = 0.5, \ldots, 1$

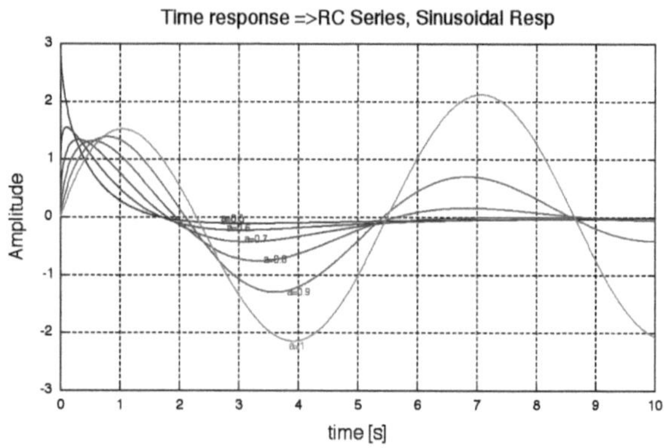

Fig. 4.8 Sinusoidal responses of the fractional RC circuit for $\alpha = 0.5, \ldots, 1$

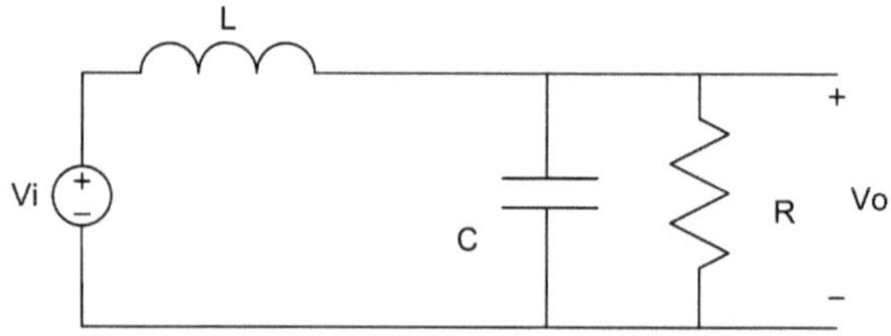

Fig. 4.9 L series R//C circuit

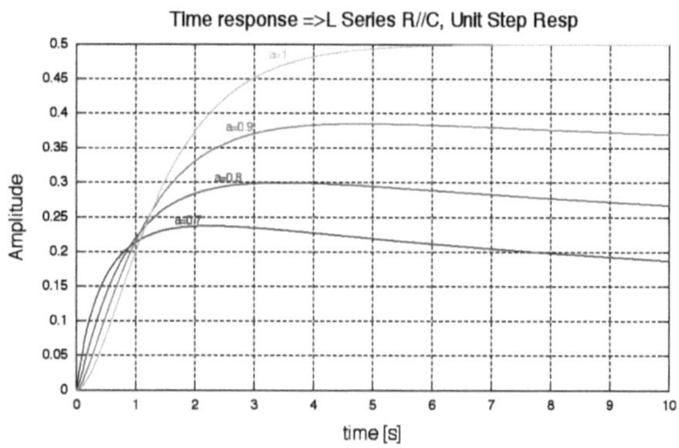

Fig. 4.10 Step responses for the L series R//C circuit

Laplace transform (ULT) and the jump formula that is a result of the distribution (generalized function) theory [20, 21].

The problems found in concrete applications have been addressed and are motivated by the ULT treatment of the origin as presented in the main text books and in the fractional case by the use of derivative definitions that impose specific initial conditions that may not be the most suitable for the problem.

In current literature we find two situations:

- People who consider the Riemann–Liouville (RL) derivative and the associated initial conditions (e.g. [22–24])

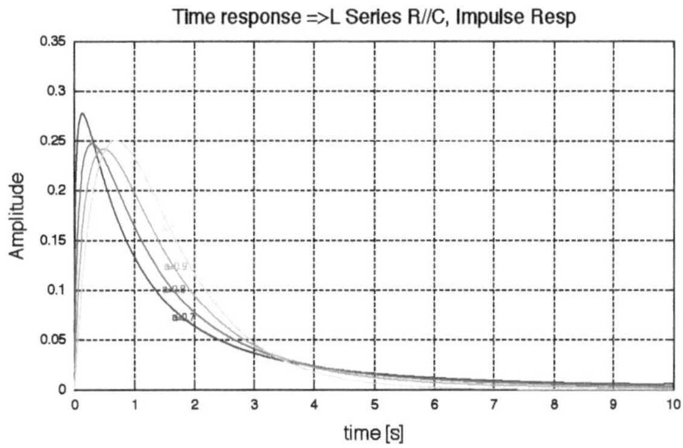

Fig. 4.11 Impulse responses for the L series R//C circuit

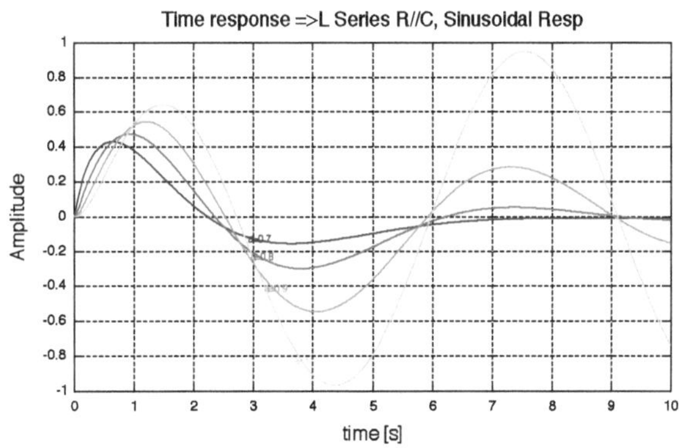

Fig. 4.12 Sinusoidal responses for the L series R//C circuit

- People that use the Caputo (C) derivative that uses integer order derivatives (e.g. [25, 26, 27].

If $x(t)$ is a causal signal and denoting the Laplace transform by LT we have for the RL case:

$$\mathrm{LT}[D^{\alpha}x(t)] = \mathrm{LT}\left[D^{m}[D^{-(m-\alpha)}x(t)]\right] = s^{\alpha}X(s) - \sum_{i=0}^{m-1} s^{m-i-1}D^{i-m+\alpha}x(0) \quad (4.27)$$

where m is the least integer greater than or equal to α. In the Caputo case, we have

$$\mathrm{LT}[D^{\alpha}x(t)] = \mathrm{LT}\left[D^{-(m-\alpha)}[D^{m}x(t)]\right] = s^{\alpha}X(s) - \sum_{i=0}^{m-1} s^{i-m+\alpha}D^{m-i-1}x(0) \quad (4.28)$$

In the last years the second approach has been favoured relatively to the first, because it is believed that the RL case leads to initial conditions without physical meaning. This was contradicted by Heymans and Podlubny [28] that studied several cases and gave physical meaning to the RL initial conditions, by introducing the concept of "inseparable twin". On the other hand,[29] shows that both types of initial conditions can appear. Similar position is assumed by Gorenflo and Mainardi [30] and Bonilla et al. [31].

In Ortigueira [32] and Ortigueira and Coito [33, 34] these positions were criticised: *the initial conditions belong to the system at hand and cannot depend on the used derivative*. The "initial conditions" of RL or C derivatives are needed to compute them correctly, but may not have any direct relation with the system initial conditions.

The problem is faced here with all the generality. The approach we are going to describe is based on the following assumptions:

- All the involved signals are defined over the whole set of real numbers.
- If the systems are observed for $t > t_0, t_0 \in R$, our observation window is the Heaviside unit step function, $u(t - t_0)$.
- The initial conditions depend on the past input and output of the system, not on the actual or future.

We will put the Riemann–Liouville and Caputo derivatives in terms of this general frame work and discover which are the equations suitable for RL and C derivatives.

4.7.2 The Initialization Problem

Let us assume that we have a fractional linear system described by the fractional differential equation (4.1):

$$\sum_{n=0}^{N} a_n D^{\gamma_n} y(t) = \sum_{m=0}^{M} b_m D^{\gamma_m} x(t), \quad \gamma_n < \gamma_{n+1} \tag{4.29}$$

where γ_n, $n = 0, 1, 2, \ldots$ are derivative orders that we will assume to be positive real numbers. This equation is valid for every $t \in R$.

As it is well known, the solution of the above equation has two terms: the forced (or evoked) and free (or spontaneous). This second term depends only on the state of the system at the reference instant that we will assume to be $t = 0$. This state constitutes or is related to the initial conditions. These are the values at $t = 0$ of variables in the system and associated with stored energy. *It is the structure of the system that imposes the initial conditions, not the eventual way of computing the derivatives*. The instant where the initial conditions are taken is very important, but it has not received much attention. In most papers, people don't care and use $t = 0$. This happens in most mathematical books and papers (see the references in Lunberg et al.). Others use $t = 0^+$, motivated by the requirement of continuity of the functions for $t = 0$ and the initial value theorem. However and as pointed out by Lunberg et al. (2007), we must retain the initial conditions at $t = 0^-$, because the initial conditions represent the past of the system and do not have any relation with the future inputs.

In problems with nonzero initial conditions it is a common practice to introduce the one-sided Laplace transform. However, there is no particular justification for such introduction. The initial conditions must appear independently of using or not a transform. In fact, we intend to solve a given differential equation (4.1) for values of t greater than a given initial instant, that, without loosing generality, we can assume to be the origin. To treat the question, it is enough to multiply both members of the equation by the unit step Heaviside function, $u(t)$, and rearrange the equation terms as shown next with a simpler example. Consider the ordinary constant coefficient differential equation:

$$y^{(N)}(t) + ay(t) = x(t), \quad N \in Z_0^+ \tag{4.30}$$

Assume that the products $y^{(N)}(t)u(t)$ and $x(t)u(t)$ can be considered as distributions and that we want to solve Eq. 4.30 for $t > 0$. The multiplication by $u(t)$ leads to

$$y^{(N)}(t)u(t) + ay(t)u(t) = x(t)u(t) \tag{4.31}$$

Thus, we have to relate $y^{(N)}(t)u(t)$ with $[y(t) \cdot u(t)]^{(N)}$. This can be done recursively provided that we account for the properties of the distribution $\delta(t)$ and its derivatives. We obtain the well known result:

$$y^{(N)}(t) \cdot u(t) = [y(t) \cdot u(t)]^{(N)} - \sum_{i=0}^{N-1} y^{(N-1-i)}(0) \cdot \delta^{(i)}(t) \tag{4.32}$$

that states that $y^{(N)}(t) \cdot u(t) = [y(t) \cdot u(t)]^{(N)}$ for $t > 0$. They are different at $t = 0$. This is the reason why we speak in initial values as being equivalent to initial conditions. In the above equation we have

$$[y(t) \cdot u(t)]^{(N)} + a[y(t) \cdot u(t)] = x(t) + \sum_{i=0}^{N-1} y^{(N-1-i)}(0) \cdot \delta^{(i)}(t) \tag{4.33}$$

The initial conditions appear naturally, without using any transform. It is interesting to remark that the second term on the right in (4.33) is constituted by the derivatives of the Heaviside functions that we are needing for making continuous the left hand function before computing the derivative. For example, $y(t)u(t)$ is not continuous at $t = 0$, but $y(t)u(t) - y(0)u(t)$ is, so, its derivative is given by $[y(t) \cdot u(t)]' - y(0) \cdot \delta(t)$. The process is repeated.

In fractional case, the problem is similar, but it is not so clear the introduction of the initial conditions, because the involved functions can be infinite at $t = 0$.

4.8 Some Facts

When computing a α order derivative, it is easy to deduce, that:

(a) Different steps lead to different initial values.
(b) In the differentiation steps some orders and corresponding initial values are fixed and defined by the equation: in the left hand side in (4.29) when "going" from 0 to v_N, we have to "pass" by all the v_i $(i = 1, \ldots N-1)$—with the corresponding initial conditions. However we can compute other derivatives of orders $\gamma_i (v_i < \gamma_i < v_{i+1})$ that introduce initial conditions too.
(c) If in (4.29) all the v_n are rational numbers, the differential equation can always be written as in (4.8):

$$\sum_{n=0}^{N} a_n D^{nv} y(t) = \sum_{m=0}^{M} b_m D^{mv} x(t)$$

leading as to conclude that the "natural" initial values will be $D^{nv} y(t)|_{t=0}$ for $n = 0, \ldots, N-1$ and $D^{nv} x(t)|_{t=0}$ for $n = 0, \ldots, M-1$.
(d) Independently of the way followed to compute a given derivative, the Laplace Transform of the derivative satisfy: $\mathrm{LT}[D^\alpha f(t)] = s^\alpha \mathrm{LT}[f(t)]$. So, the different steps in the derivative computation correspond to different decompositions of the number α:

$$\alpha = \sum_i \sigma_i \tag{4.34}$$

These considerations lead us to conclude that the initial condition problem in the fractional case has infinite solutions.

4.9 An Example

In practical applications we can find several examples of systems with Transfer Functions given by:

$$H(s) = \frac{Q}{s^\alpha}$$

where Q is a constant and $-1 < \alpha < 1$. They are known as "constant phase elements" [35, 36]. In particular, the supercapacitors are very important. The capacitor case is well studied by Westerlund [37], where he shows that the "natural" initial value is the voltage at $t = 0$ that influences the output of the system through an initial function proportional to $t^{-\alpha} u(t)$.

With this example we had in mind to remark that the structure of the problem may lead us to decide what initial condition we should use—it is an engineering problem, not mathematical.

4.9.1 The Initial-Value Theorem

The Abelian initial value theorem [38] is a very important result in dealing with the Laplace Transform. This theorem relates the asymptotic behaviour of a causal signal, $\varphi(t)$, as $t \to 0+$ to the asymptotic behaviour of its Laplace Transform, $\Phi(\sigma) = \text{LT}[\varphi(t)]$, as $\sigma = \text{Re}(s) \to +\infty$.

The initial-value theorem—assume that $\varphi(t)$ is a causal signal such that in some neighbourhood of the origin is a regular distribution corresponding to an integrable function. Also, assume that there is a real number $\beta > -1$ such that $\lim_{t\to 0+} \frac{\varphi(t)}{t^\beta}$ exists and is a finite complex value. Then

$$\lim_{t\to 0+} \frac{\varphi(t)}{t^\beta} = \lim_{\sigma\to\infty} \frac{\sigma^{\beta+1}\Phi(\sigma)}{\Gamma(\beta+1)} \tag{4.35}$$

For proof see [38]. Let $-1<\alpha<\beta$. Then

$$\lim_{t\to 0+} \frac{\varphi(t)}{t^\alpha} = \lim_{t\to 0+} \frac{\varphi(t)}{t^\beta}\frac{t^\beta}{t^\alpha} = 0 \tag{4.36}$$

because the first factor has a finite limit given in (4.35) and the second zero as limit. Similarly, if $\beta<\alpha$,

$$\lim_{t\to 0+} \frac{\varphi(t)}{t^\alpha} = \infty \tag{4.37}$$

This suggests us that, near $t = 0$, $\varphi(t)$ must have the format: $\varphi(t) = s(t) \cdot t^\beta u(t)$, where $s(t)$ is regular at $t = 0$.

We may wander about a possible contradiction between the initial value theorem and the considerations we did before concerning the initial conditions of the systems. There is no contradiction for several reasons. Remark that the initial value theorem does not tell us how to select the initial conditions. It states a relation between the "starting" value of a given causal signal and the "final" value of its Laplace transform. On the other hand, the "local value" of a given generalised function and also of a fractional derivative depends on all the history of the function not only what happens at a given point. This means that a signal cannot jump from a value $x(0^-)$ to a different $x(0^+)$ without any input. So unless necessary, we will work with $x(0)$.

4.10 A Solution for the Initial Value Problem

4.10.1 The Watson–Doetsch Class

Let us consider the class of functions with Laplace Transform analytic for $\text{Re}(s) > \gamma$. To the subclass of functions such that

$$\varphi(t) \approx t^{\beta} \cdot \sum_{n=0}^{\infty} a_n \frac{t^{n\nu} u(t)}{\Gamma(\beta + 1 + n\nu)} \tag{4.38}$$

as $t \to 0+$ where $\beta > -1$ and $\nu > 0$. The powers are defined in the principal branch. For our applications to differential equations, we will assume that ν is greater than the maximum derivative order. The Watson–Doetsch lemma [17], states that the LT $\Phi(s)$ of $\varphi(t)$ satisfies:

$$\Phi(s) \approx \frac{1}{s^{\beta+1}} \sum_{n=0}^{\infty} a_n \frac{1}{s^{n\nu}} \tag{4.39}$$

As $s \to \infty$ and $\mathrm{Re}(s) > 0$.

As it is clear, these functions verify the initial value theorem. On the other hand, $\varphi(t)$ in (4.38) has a format very common in solving the fractional differential equations as we saw before.

For this reason, we will use "=" instead of "≈" in the following. On the other hand, as $\sigma^{\beta}\Phi(\sigma) = \mathrm{LT}\left[D^{\beta}\varphi(t)\right]$,

$$\lim_{\sigma \to \infty} \sigma\left[\sigma^{\beta}\Phi(\sigma)\right] = D^{\beta}\varphi(t)\big|_{t=0+} \tag{4.40}$$

by the usual initial value theorem. So,

$$D^{\beta}\varphi(t)\big|_{t=0+} = \lim_{\sigma \to \infty} \sigma^{\beta+1}\Phi(\sigma) \tag{4.41}$$

that is a generalisation of the usual initial value theorem, obtained when $\beta = 0$. Here, we remind that the impulse response of the differintegrator, $\delta^{(\alpha)}(t) = \mathrm{LT}^{-1}[s^{\alpha}]$, given by:

$$\delta^{(\alpha)}(t) = \begin{cases} \frac{t^{-\alpha-1}}{\Gamma(-\alpha)} u(t) & \alpha \in R - Z^{+} \\ \delta^{(n)}(t) & \alpha \in Z^{+} \end{cases}$$

Because $u(t) = \frac{D^{\beta}\left[t^{\beta}u(t)\right]}{\Gamma(\beta+1)}$ and using (4.40) and (4.41), we obtain:

$$\lim_{t \to 0+} \frac{\varphi(t)}{t^{\beta}} = \lim_{t \to 0^{+}} \frac{D^{\beta}\varphi(t)}{D^{\beta}\left[t^{\beta}u(t)\right]} = \frac{\varphi^{(\beta)}(0+)}{\Gamma(\beta+1)} \tag{4.42}$$

that is very similar to the usual l'Hôpital rule used to solve the 0/0 problems.

Now, let us assume that $\varphi(t)$ is written as:

$$\varphi(t) = t^{\beta} \cdot f(t) \cdot u(t) \tag{4.43}$$

where $f(t)$ is given by:

$$f(t) = \sum_{n=0}^{\infty} a_n \frac{t^{n\nu} u(t)}{\Gamma(\beta + 1 + n\nu)} \tag{4.44}$$

Attending to Eqs. 4.40 to 4.42, it is not hard to conclude that, when $t \to 0+$, we have:

$$D^{\alpha}\varphi(t)|_{t=0+} = \begin{cases} 0 & \text{if } \alpha < \beta \\ f(0+)\Gamma(\beta+1) & \text{if } \alpha = \beta \\ \infty & \text{if } \alpha > \beta \end{cases} \tag{4.45}$$

All the derivatives of order $\alpha < \beta$ have a zero initial value, while all the derivatives of order greater than β are infinite at $t = 0$. To obtain a continuous function we have to remove a term proportional to $t^{\beta-\alpha}u(t)$. This is important in dealing with differential equations and will be done in the following solution. Return back to Eq. 4.29. The previous considerations lead us to state for $y(t)$—and similarly for $x(t)$—the following format:

$$y(t) = \sum_{k=0}^{N} f_n(t)t^{\gamma_n}u(t) \tag{4.46}$$

where $0 < \gamma_n < \gamma_{n+1}$—according to the initial value theorem, we could use $-1 < \gamma_n$, but in our present application it does not interest. N is a positive integer that may be infinite, and the functions $f_n(t)$ ($n = 0, \ldots, N$) and their derivatives of orders less than or equal to γ_N are assumed to be regular at $t = 0$. We may assume them to have the format given by (4.46) and verifying (4.45).

4.10.2 Step by Step Differentiation

It is interesting to see how the initial values appear and their meaning. Let $y(t)$ be a signal given by (4.46). Let us introduce a sequence β_n by:

$$\beta_n = \gamma_n - \sum_{k=0}^{n-1} \beta_k, \quad \beta_0 = \gamma_0 \tag{4.47}$$

Let us see what happens proceeding step by step.

(a) According to our assumptions β_0 is the least real for which $\lim_{t\to 0} \frac{y(t)}{t^{\beta_0}}$ is finite and nonzero. Let it be $\frac{y^{(\beta_0)}(0)}{\Gamma(\beta_0+1)}$. All the derivatives $D^{\alpha}y(t)(\alpha < \beta_0)$ are continuous at $t = 0$ and assume a zero value. The β_0 order derivative assumes the value $y^{(\beta_0)}(0)$ and we can construct the function

$$\varphi^{(\beta_0)}(t) = [y(t) \cdot u(t)]^{(\beta_0)} - y^{(\beta_0)}(0)u(t) \tag{4.48}$$

that is continuous and assumes a zero value at $t = 0$.

(b) Now, β_1 is the least real for which $\lim_{t\to 0} \frac{\varphi^{(\beta_0)}(t)}{t^{\beta_1}}$ is finite and nonzero. Let it be $\frac{y^{(\beta_0+\beta_1)}(0)}{\Gamma(\beta_1+1)}$. Thus β_1 derivative of $\varphi^{(\beta_0)}(t)$ is given by:

$$\varphi^{(\beta_0+\beta_1)}(t) = [y(t) \cdot u(t)]^{(\beta_0+\beta_1)} - y^{(\beta_0)}(0)\delta^{(\beta_1-1)}(t) - y^{(\beta_0+\beta_1)}(0)u(t) \quad (4.49)$$

is again continuous at $t = 0$.

(c) Again β_2 is the least real for which $\lim_{t\to 0} \frac{\varphi^{(\beta_0+\beta_1)}(t)}{t^{\beta_2}}$ is finite and nonzero. Let it be $\frac{y^{(\beta_0+\beta_1+\beta_2)}(0)}{\Gamma(\beta_2+1)}$. Thus

$$f^{(\beta_0+\beta_1+\beta_2)}(t) = [y(t).u(t)]^{(\beta_0+\beta_1+\beta_2)} - y^{(\beta_0)}(0)\delta^{(\beta_1+\beta_2-1)}(t) - y^{(\beta_0+\beta_1)}(0)\delta^{(\beta_2-1)}(t) - y^{(\beta_0+\beta_1+\beta_2)}(0)u(t) \quad (4.50)$$

is again continuous at $t = 0$.

(d) Continuing with this procedure, we obtain a function:

$$\varphi^{(\gamma_N)}(t) = [y(t).u(t)]^{(\gamma_N)} - \sum_0^{N-1} y^{(\gamma_m)}(0)\delta^{(\gamma_N-\gamma_i-1)}(t) \quad (4.51)$$

that is not continuous at $t = 0$, but it can be made continuous if we subtract it $y^{(\gamma_N)}(0)u(t)$. Equation 4.51 states the general formulation of the initial value problem solution. As we can see, the initial values prolong their action for every $t > 0$. This means that we have a memory about the initial conditions that decreases very slowly. Using the LT, we obtain:

$$LT[\varphi^{(\gamma_N)}(t)] = s^{\gamma_N}Y(s) - s^{\gamma_N}\sum_0^{N-1} y^{(\gamma_m)}(0)s^{-\gamma_i-1} \quad (4.52)$$

that is a generalization of the usual formula for introducing the initial conditions. Using this procedure in both members of Eq. 4.29 leads to the initial condition complete equation

$$\sum_{i=0}^{N} a_i \cdot [y(t) \cdot u(t)]^{(\gamma_i)} = \sum_{i=0}^{M} b_i \cdot [x(t) \cdot u(t)]^{(\gamma_i)} + \sum_{i=1}^{N} a_i \cdot \sum_0^{i-1} y^{(\gamma_m)}(0)\delta^{(\gamma_i-\gamma_m-1)}(t) - \sum_{i=1}^{M} b_i \sum_0^{i-1} x^{(\gamma_m)}(0)\delta^{(\gamma_i-\gamma_m-1)}(t) \quad (4.53)$$

Equation 4.53 states a general formulation of the initial value problem solution.

4.10.3 Examples

Consider the system described by Eq. 4.30 with $\alpha = 3/2$. As in the equation we only have two terms we are not constrained and can choose any “way” to go from 0 to α. We are going to consider four cases:1 $- \gamma_i = 3/2 \cdot i(i = 0, 1)$ or $\beta_0 = 0$ and $\beta_1 = 3/2$. From (4.52), we have

$$\mathrm{LT}[\varphi^{(3/2)}(t)] = s^{3/2}Y(s) - y(0)s^{1/2} \tag{4.54}$$

The free term is then:

$$\Phi_f(s) = y(0) \cdot \frac{s^{1/2}}{s^{3/2} + a} \tag{4.55}$$

$2 - \gamma_i = 1/2 \cdot i \ (i = 0, 1, 2, 3)$ or $\beta_0 = 0$ and $\beta_i = 1/2 \ (i = 1, 2, 3)$. We have now:

$$\mathrm{LT}[\varphi^{(3/2)}(t)] = s^{3/2}Y(s) - \sum_0^2 y^{(m/2)}(0)s^{(3-m)/2-1} \tag{4.56}$$

with

$$\Phi_f(s) = \frac{\sum_0^2 y^{(m/2)}(0)s^{(3-m)/2-1}}{s^{3/2} + a} \tag{4.57}$$

as the corresponding free term.

$3 - \gamma_i = 1/2 + i \quad (i = 0, 1)$ or $\beta_0 = 1/2$ and $\beta_1 = 3/2$, giving the Riemann–Liouville solution:

$$\mathrm{LT}[\varphi^{(3/2)}(t)] = s^{3/2}Y(s) - y^{(1/2)}(0) \tag{4.58}$$

The same solution can be obtained with $\gamma_l = -1/2 + i \quad (i = 0, 1, 2)$. Now, the free term is given by:

$$\Phi_f(s) = y^{(1/2)}(0) \cdot \frac{1}{s^{3/2} + a} \tag{4.59}$$

$4 - \gamma_i = i \quad (i = 0, 1)$ and $\gamma_2 = 2 - 1/2$. It comes:

$$\mathrm{LT}[\varphi^{(3/2)}(t)] = s^{3/2}Y(s) - \sum_0^1 y^{(m)}(0)s^{3/2-m-1} \tag{4.60}$$

giving the free term:

$$\Phi_f(s) = \frac{\sum_0^1 y^{(m)}(0)s^{(3/2-m-1)}}{s^{3/2} + a} \tag{4.61}$$

The situation is somehow different if we have an intermediary term as it is the case of the equation:

$$y^{(\alpha)}(t) + ay^{(1)}(t) + by(t) = x(t) \tag{4.62}$$

Now, when going from $\gamma = 0$ to $\gamma = 3/2$, we have to "pass" by $\gamma = 1$. Obviously, we can force the corresponding initial value to be zero.

It is interesting to see what happens when we consider an ordinary integer order differential equation as a special case of a fractional differential equation. Consider the simple case:

$$y'(t) + ay(t) = x(t) \tag{4.63}$$

Putting $\gamma_i = 1/2i$ $(i = 0, 1, 2)$, we have

$$f'(t) = [y(t) \cdot u(t)]' - \sum_{i=0}^{1} y^{(1/2.i_i,)}(0)\delta^{(-1/2.i_i)}(t) \tag{4.64}$$

leading to a free term with LT given by:

$$F_f(s) = \frac{y(0) + y^{(1/2)}(0)s^{-1/2}}{s+a} \tag{4.65}$$

Obviously different from the usual that we obtain by putting $y^{(1/2)}(0) = 0$.

4.10.4 Special Cases

4.10.4.1 Riemann–Liouville

The left Riemann–Liouville fractional derivative as it is commonly presented can be represented by the following double convolution as seen in Chap. 2

$$f_{RL}^{(\alpha)}(t) = \delta_+^{(n)}(t) * \left\{ f(t) * \delta_+^{(-\nu)}(t) \right\}$$

where $\alpha = n - \nu$, $\delta_+^{(n)}(t)$ is the nth derivative of the Dirac impulse, and

$$\delta_+^{(-\nu)}(t) = \frac{t^{\nu-1}}{\Gamma(\nu)}u(t), \; 0<\nu<1$$

In terms of the operator D, we can write:

$$f_{RL}^{(\alpha)}(t) = D\{D[D\ldots D^{-\nu}]\}f(t)$$

So, we have an anti-derivative followed by a sequence of N order one derivatives. This leads to $\beta_0 = \gamma = -\nu$ and $\beta_i = 1$, and $\gamma_i = \gamma + i$, for $i = 1, \ldots, N$. Then,

$$\varphi^{(N+\gamma)}(t) = [y(t) \cdot u(t)]^{(N+\gamma)} - \sum_{0}^{N-1} y^{(m+\gamma)}(0)\delta^{(N-1-m)}(t) \tag{4.66}$$

and

$$\mathrm{LT}[\varphi^{(N+\gamma)}(t)] = s^{N+\gamma}Y(s) - \sum_{0}^{N-1} y^{(m+\gamma)}(+)s^{N-m-1} \tag{4.67}$$

With $\alpha = N + \gamma$, this relation can be rewritten as:

$$\mathrm{LT}[\varphi^{(\alpha)}(t)] = s^{\alpha}Y(s) - \sum_{0}^{N-1} y^{(\alpha-1-i)}(0)s^{i} \tag{4.68}$$

that is the current Riemann–Liouville solution. With the above set of orders, we obtain for the initial condition complete equation

$$\sum_{n=0}^{N} a_n D^{\gamma+n} y(t) = \sum_{m=0}^{M} b_m D^{\gamma+m} x(t) + \sum_{i=1}^{N} a_i \cdot \sum_{0}^{i-1} y^{(\gamma+m)}(0)\delta^{(i-m-1)}(t)$$
$$- \sum_{i=1}^{M} b_i \sum_{0}^{i=1} x^{(\gamma+m)}(0)\delta^{(i-m-1)}(t) \tag{4.69}$$

From this result, we immediately conclude that the RL initial conditions are suitable for solving equations of the following format:

$$\sum_{n=0}^{N} a_n D^{\gamma+n} y(t) = \sum_{m=0}^{M} b_m D^{\gamma+m} x(t) \tag{4.70}$$

that is a very restrict class.

4.10.4.2 Caputo

Similarly to the RL case, the left Caputo fractional derivative as it is commonly presented can be represented by the following double convolution:

$$f_C^{(\alpha)}(t) = \left\{ f(t) * \delta_{+}^{(N)}(t) \right\} * \delta_{+}^{(-\nu)}(t)$$

In terms of the operator D, we can write:

$$f_C^{(\alpha)}(t) = D^{-\nu}\{D[D\ldots D]\}f(t)$$

corresponding to a sequence of N order one derivatives and an integration. The Caputo case is not in the framework considered in Sect. 4.10.2. In fact, we considered there that the γ_n $(n = 0, \ldots, N)$ form an increasing sequence. In Caputo differentiation, we have $\gamma_n = n$ for $(n = 0, \ldots, N-1)$ and $\gamma_N = N - \varepsilon$ with $0 < \varepsilon < 1$. However, the anti-derivative does not introduce nonzero initial conditions, we have:

$$\varphi^{(\gamma_N)}(t) = [y(t) \cdot u(t)]^{(\gamma_N)} - \sum_{0}^{N} y^{(i)}(0)\delta^{(N-i-1-\varepsilon)}(t) \tag{4.71}$$

or, putting $\alpha = N - \varepsilon$,

$$\varphi^{(\alpha)}(t) = [y(t) \cdot u(t)]^{(\alpha)} - \sum_{0}^{N} y^{(i)}(0)\delta^{(\alpha-i-1)}(t) \tag{4.72}$$

that is the usual way of presenting the C derivative. With this result and following a procedure similar to the one used in the RL case, we can write:

$$\begin{aligned} D^{N-\varepsilon}y(t) + \sum_{n=0}^{N-1} a_n D^n y(t) = b_0 D^{M-\varepsilon}x(t) + \sum_{m=0}^{M-1} b_m D^n x(t) \\ + \sum_{i=1}^{N} a_i \cdot \sum_{0}^{i-1} y^{(j)}(0)\delta^{(N-j-1-\varepsilon)}(t) \\ - \sum_{i=1}^{M} b_i \sum_{0}^{i-1} x^{(j)}(0)\delta^{(N-j-1-\varepsilon)}(t) \end{aligned} \tag{4.73}$$

So and as in the RL case, the C derivative is suitable for dealing with equations with the general format:

$$D^{N-\varepsilon}y(t) + \sum_{n=0}^{N-1} a_n D^n y(t) = b_0 D^{M-\varepsilon}x(t) + \sum_{m=0}^{M-1} b_m D^n x(t) \tag{4.74}$$

that represents again a very restrict class of systems.

4.10.4.3 The Rational Order Case

If all the orders in (4.29) are rational we can always put them as multiple of a given rational γ: $\gamma_i = i\gamma$, for $i = 0, 1, \ldots, N$. We have: $\beta_0 = 0, \beta_i = \gamma$, for $i = 1, \ldots, N-1$. Then, (4.51) will be transformed into

$$\varphi^{(n\gamma)}(t) = [y(t) \cdot u(t)]^{(n\gamma)} - \sum_{0}^{n-1} y^{(m\gamma)}(0)\delta^{(n-i)\gamma-1}(t) \tag{4.75}$$

that inserted in (4.29), gives

$$\begin{aligned} \sum_{i=0}^{N} a_i \cdot [y(t) \cdot u(t)]^{(i\gamma)} = \sum_{i=0}^{M} b_i \cdot [x(t) \cdot u(t)]|^{(i\gamma)} + \sum_{i=1}^{N} a_i \cdot \sum_{j=0}^{i-1} y^{(i\gamma)}(0) \cdot \delta^{((i-j)\gamma-1)}(t) \\ - \sum_{i=1}^{M} b_i \sum_{j=0}^{i-1} x^{(j\gamma)}(0) \cdot \delta^{((i-j)\gamma-1)}(t) \end{aligned} \tag{4.76}$$

This is also valid even if γ is not rational. This equation can be solved using the two-sided LT.

4.11 State-Space Formulation

In some applications, e.g. control, the state-space formulation is very important. It is not hard to obtain it from (4.8). It can be written for the time-variant case as:

$$s^{(v)}(t) = A(t) \cdot s(t) + B(t) \cdot x(t) \tag{4.77}$$

$$y(t) = C(t) \cdot s(t) + D(t) \cdot x(t) \tag{4.78}$$

To solve the dynamic equation, it is necessary to introduce the *fractional state transition operator* $\Phi(t, \tau)$, which is a generalisation of the usual state transition operator. Heuristically, we could conclude that the required operator can be represented by the usual Peano–Baker series with a substitution of an v-order integration for the usual one. In the time-invariant case, this operator is related to the Mittag–Leffler function. However, it is very difficult to manipulate. Besides it does not enjoy all the features of the ordinary one, namely, the semi-group property even in the time-invariant case.[5] This fact has a very important consequence: the operator $\Phi(t, \tau)$ is not the inverse operator of $\Phi(t, \tau)$. Such inverse operator will be obtained probably with the help of the anti-causal differintegration operator. This must be a subject of further research.

4.12 Conclusions

In this chapter, we presented a new class of linear systems: the fractional continuous-time linear systems. The results obtained in Chap. 1 allowed us to present an approach that is very similar to the one used in the study of the ordinary linear systems, namely we were led to the notions of fractional impulse and frequency responses. We showed how to compute them. The initial condition problem was treated with generality. We made also a brief study of the stability of these systems and introduced the state-space representation.

References

1. Koh CG, Kelly JM (1990) Application of fractional derivatives to seismic analysis of basedisolated models. Earthq Eng Struct D 19:229–241
2. Anastasio TJ (1994) The fractional-order dynamics of brainstem vestibulo-ocular dynamics of brainstem vestibular-oculomotor neurons. Biol Cybern 72:69–79

---

[5] In fact $f(0,t) = \int_0^t (t-\tau)^{-\alpha-1} \mathrm{d}\tau = \int_0^{t_1} (t-\tau)^{-\alpha-1} \mathrm{d}\tau + \int_{t_1}^t (t-\tau)^{-\alpha-1} \mathrm{d}\tau \neq f(0,t_1) + f(t_1,t)$ $\Rightarrow \Phi(0,t_1) \cdot \Phi(t_1,t)$.

3. Gaul L, Klein P, Kemple S (1991) Damping description involving fractional operators. Mech Syst Signal Process 5(2):81–88
4. Makris N, Constantinou MC (1991) Fractional-derivative Maxwell model for viscous dampers. J Struct Eng 117(9):2708–2724
5. Clerc JP, Tremblay A-MS, Albinet G, Mitescu, CD (1984) a. c. Response of Fractal Networks. Le Journal de Physique-Lettres 45(19)
6. Oustaloup A (1981) Fractional order sinusoidal oscillators: optimization and their use in highly linear FM modulation. IEEE Trans Circuits Syst 28(10):1007–1009
7. Matignon D, d' Andréa-Novel B (1995) Décomposition modale fractionaire de ĺ équation des ondes avec pertes viscothermiques, Tech. Rep. 95 C 001, École Nationale Supérieure des Télécommunications
8. Oustaloup A, Mathieu B, Lanusse P (1995) The CRONE control of resonant plants: application to a flexible transmission. Eur J Control 1(2):113–121
9. Sabatier JM, Agrawal OP, Machado JAT (2007) Advances in fractional calculus: theoretical developments and applications in physics and engineering. Springer, Berlin
10. Machado JAT, Azenha A (1998) Position/Force fractional control of mechanical manipulators. In: Proceedings of the 1998 5th international workshop on advanced motion control, Coimbra, Portugal, pp. 216–221
11. Magin RL (2006) Fractional calculus in bioengineering. Begell House, Connecticut
12. Craiem D, Armentano RL (2007) A fractional derivative model to describe arterial viscoelasticity. Biorheology 44:251–263
13. Kolmogorov NA (1940) Wienersche Spiralen und Einige Andere Interessante Kurven im Hilbertschen Raum. Compte Rendus de l'Académie des Sciences de l'Union des Républiques Soviétiques Socialistes 26(2):115–118
14. Mandelbrot BB, Van Ness JW (1968) The fractional Brownian motions, fractional noises and applications. SIAM Rev 10(4):422–437
15. Ortigueira MD, Batista AG (2008) On the relation between the fractional Brownian motion and the fractional derivatives. Phys Lett A 372(7):958–968
16. Ortigueira MD (2000) Introduction to fractional signal processing. Part 1: continuous-time systems. In: IEE proceedings on vision, image and signal processing, no 1, February, pp 62–70
17. Henrici P (1991) Applied and computational complex analysis, vol 2. Wiley, New York, pp 389–391
18. Kilbas AA, Srivastava HM, Trujillo JJ (2006) Theory and applications of fractional differential equations. Elsevier, Amsterdam
19. Lundberg HK, Miller HR, Trumper DL (2007) Initial conditions, generalized functions, and the Laplace transform: Troubles at the origin. IEEE Control Sys Mag pp. 22–35
20. Ferreira JC (1997) Introduction to the theory of distributions. Pitman monographs and surveys in pure and applied mathematics, July
21. Hoskins RF (1999). Delta Functions. Horwood Series in Mathematics & Applications. Chichester, England
22. Samko SG, Kilbas AA, Marichev OI (1987) Fractional integrals and derivatives—theory and applications. Gordon and Breach Science Publishers, New York
23. Miller KS, Ross B (1993) An introduction to the fractional calculus and fractional differential equations. Wiley, New York
24. Podlubny I (1999) Fractional differential equations: an introduction to fractional derivatives, fractional differential equations, to methods of their solution and some of their applications. Academic Press, San Diego
25. Davison M, Essex C (1998) Fractional differential equations and initial value problems. Math. Scil 23:108–116
26. Mainardi F, Pagnini G (2003). The Wright functions as solutions of the time-fractional diffusion equation. JAMC 141:(1) 51–62
27. Daftardar-Gejji V, Jafari H (2006) An iterative method for solving nonlinear functional equations. J Math Anal Appl 316:753–763

28. Heymans N, Podlubny I (2005) Physical interpretation of initial conditions for fractional differential equations with Riemann–Liouville fractional derivatives. Rheol Acta 37:1–7
29. Agrawal O (2006) Fractional variational calculus and the transversality conditions. J Phys A: Math Gen 39:10375–10384
30. Gorenflo R, Mainardi F (1997) Integral and differential equations of fractional order. In: Carpinteri A, Mainardi F (eds) Fractals and fractional calculus in continuum mechanics. Springer Verlag, New York, pp 223–276
31. Bonilla B, Rivero M, Trujillo JJ (2007) On systems of linear fractional differential equations with constant coefficients. Appl Math Comp 187(9):68–78
32. Ortigueira MD (2003) On the initial conditions in continuous-time fractional linear systems. Signal Processing 83(11):2301–2309
33. Ortigueira MD, Coito FJ (2010) System initial conditions vs Derivative initial conditions, computers and mathematics with applications. Special issue on fractional differentiation and Its applications, Vol. 59(5):pp. 1782–1789
34. Ortigueira MD, Coito FJ (2008) Initial conditions: what are we talking about?. In: Proceedings of the 3rd IFAC workshop on fractional differentiation and its applications. Ankara, Turkey, 5–7 Nov 2008
35. Mahon PJ, Paul GL, Kesshishian SM, Vassallo AM (2000) Measurement and modelling of high-power performance of carbon-based supercapacitors. J Power Sources 91(1):68–76
36. Itagaki M, Taya A, Watanabe K, Noda K (2002) Deviations of capacitive and inductive loops in the electrochemical impedance of a dissolving iron electrode. Anal Sci 18(6):641–644
37. Westerlund S (2002) Dead matter has memory. Causal Consulting, Kalmar
38. Zemanian AH (1987) Distribution theory and transform analysis. Dover Publications, New York

Chapter 5
Two-Sided Fractional Derivatives

5.1 Motivation

In previous chapters the causal and anti-causal fractional derivatives were presented. An application to shift-invariant linear systems was studied. Those derivatives were introduced into four steps:

1. Use as starting point the Grünwald–Letnikov differences and derivatives.
2. With an integral formulation for the fractional differences and using the asymptotic properties of the Gamma function obtain the generalised Cauchy derivative.
3. The computation of the integral defining the generalised Cauchy derivative is done with the Hankel path to obtain regularised fractional derivatives.
4. The application of these regularised derivatives to functions with Laplace transform, we obtain the Liouville fractional derivative and from this the Riemann–Liouville and Caputo, two-step derivatives.

Here we will repeat the procedures for the centred (two-sided) derivatives. As we enhanced in Chap. 2, the GL derivative and those obtained from it impose preferable directions of the independent variable. We said there that the forward derivative was causal. However, there are many physical space dependent phenomena without any privilegiate direction. This means that we need a derivative suitable for these situations. To motivate the appearance of another derivative we are going to consider the following problem: which is the autocorrelation of the output of a fractional differintegrator when the input is white noise?

Assume that $x(t)$ is a stationary white noise process with $\sigma^2\delta(t)$ as its autocorrelation function. The autocorrelation of the output is given by:

$$R_X^\alpha(t_1,t_2) = \lim_{h\to 0+} \frac{\sum_{k=0}^{\infty}\sum_{n=0}^{\infty}\binom{\alpha}{k}(-1)^{k-n}\binom{\alpha}{n}R_f[t_1-t_2-(k-n)h]}{h^{2\alpha}}$$

M. D. Ortigueira, *Fractional Calculus for Scientists and Engineers*, Lecture Notes in Electrical Engineering, 84, DOI: 10.1007/978-94-007-0747-4_5,

With a change in the summation variable, it is not hard to show that,

$$R_X^\alpha(t_1,t_2) = R_X^\alpha(t_1 - t_2) = \sigma^2 \lim_{h\to 0+} \frac{\sum_{k=-\infty}^{\infty} R_\alpha(n)\delta[t_1 - t_2 - nh]}{h^{2\alpha}} \tag{5.1}$$

where $R_\alpha(n)$ is the discrete autocorrelation of the binomial coefficient sequence

$$R_\alpha(n) = \sum_{i=0}^{\infty} h_i \cdot h_{i+n} \tag{5.2}$$

with

$$h_n = (-1)^n \binom{\alpha}{n} u_n \tag{5.3}$$

The computation of its autocorrelation function is slightly involved. Inserting (5.3) into (5.2), we obtain

$$R_\alpha(n) = \sum_{i=0}^{\infty} (-1)^i \binom{\alpha}{i} (-1)^{i+k} \binom{\alpha}{i+n} \quad n \geq 0 \tag{5.4}$$

or

$$R_\alpha(n) = (-1)^k \sum_{i=0}^{\infty} \binom{\alpha}{i} \binom{\alpha}{i+n} \quad n \geq 0 \tag{5.5}$$

Let us introduce the Gauss Hypergeometric function [1]

$${}_2F_1(a,b,c,z) = \sum_{k=0}^{\infty} \frac{(a)_k (b)_k}{(c_k)} \frac{z^k}{\cdot k!} \tag{5.6}$$

where $c \neq 0, -1, -2, \ldots$ and $(a)_k$ is the Pochhammer symbol. The series (5.6) is convergent for $|z| \leq 1$, if $c - a - b > 0$.[1]

As:

$$\binom{\alpha}{i} = \frac{(-1)^i (-\alpha)_i}{i!} \tag{5.7}$$

and attending to

$$(i+k)! = (i+1)_k i! \tag{5.8}$$

and

[1] If $0 < b < c$ and $|\arg(1 - z)| < \pi$, that function can be represented by the Euler integral: ${}_2F_1(a,b;c;z) = \frac{\Gamma(c)}{\Gamma(b)\cdot\Gamma(c-b)} \int_0^1 t^{b-1}(1-t)^{c-b-1}(1-zt)^{-a} dt$

$$(-\alpha)_{i+k} = (-\alpha)_i(-\alpha+i)_k \tag{5.9}$$

we obtain

$$R_\alpha(n) = (-1)^n \cdot \binom{\alpha}{n} \cdot {}_2F_1(\alpha, -\alpha+n, n+1, 1) \quad n \geq 0 \tag{5.10}$$

Using the Gauss relation:

$${}_2F_1(a,b,c,1) = \frac{\Gamma(c)\Gamma(c-a-b)}{\Gamma(c-a)\Gamma(c-b)} \quad c-a-b>0 \tag{5.11}$$

we obtain after some simple manipulations:

$$R_\alpha(n) = (-1)^k \frac{\Gamma(1+2\alpha)}{\Gamma(\alpha+n+1)\Gamma(\alpha-n+1)} \tag{5.12}$$

that is an even function as expected. For $n = 0$, we obtain the power:

$$P_\alpha = \frac{\Gamma(1+2\alpha)}{[\Gamma(\alpha+1)]^2} \tag{5.13}$$

that is positive if $1 + 2\alpha > 0$, or $\alpha > -1/2$. This means that, only for those values, we may be led to a stationary stochastic process. If $\alpha = -1/2$, the process has an infinite power and can be considered as wide sense stationary. With this procedure we arrive at:

$$R_X^\alpha(\tau) = \lim_{h\to 0} \frac{\Gamma(2\alpha+1)}{h^{2\alpha}} \sum_{k=-\infty}^{+\infty} \frac{(-1)^k}{\Gamma(\alpha-k+1)\Gamma(\alpha+k+1)} \delta(\tau-kh) \tag{5.14}$$

So, the autocorrelation function of the forward α-order derivative of white noise suggests us the introduction of a new (centred) derivative similar to the GL derivative but that is two-sided in the sense of using past and future. We will proceed accordingly to the following steps:

1. Introduction of the general framework for the central (two-sided) differences, considering two cases that we will be called type 1 and type 2 differences. These are generalisations of the usual central differences for even and odd positive orders respectively.
2. Limit computation as in the usual Grünwald–Letnikov derivatives.
3. For those differences, suitable integral representations were introduced. From these representations we can obtain the derivative integral formulations by using the properties of the Gamma function. The integration is performed over two infinite lines that "close at infinite" to form a closed path. Two generalisations of the usual Cauchy derivative definition are obtained that agree with it when α is an even or an odd positive integer, respectively.
4. The computation of those integrals over a two straight lines path leads to generalisations of the Riesz potentials.

5. The most interesting feature of the obtained relations lies in the summation formulae for the Riesz potentials.

We will test the coherence of the proposed framework by applying them to the complex exponential. The results show that they are suitable for functions with Fourier transform. The formulation agrees also with Okikiolu [2] studies. Special cases are studied and some properties presented.

5.2 Integer Order Two-Sided Differences and Derivatives

We introduce Δ_c as finite two-sided (centred) difference defined by

$$\Delta_c f(t) = f(t + h/2) - f(t - h/2) \tag{5.15}$$

By repeated application, we have:

$$\Delta_e^N f(z) = \sum_{k=-N/2}^{N/2} (-1)^{N/2-k} \frac{N!}{(N/2+k)!(N/2-k)!} f(t-kh) \tag{5.16}$$

when N is even, and

$$\Delta_o^N f(z) = \sum_{k=-N/2}^{N/2^*} (-1)^{N/2-k} \frac{N!}{(N/2+k)!(N/2-k)!} f(t-kh) \tag{5.17}$$

if N is odd and where the $\sum_{k=-N/2}^{N/2^*}$ means that the summation is done over half-integer values. Using the Gamma function, we can rewrite the above formulae in the format stated as follows.

Definition 5.1 Let N be a positive even integer. We define a centred difference by:

$$\Delta_e^N f(t) = (-1)^{N/2} \sum_{k=-N/2}^{N/2} (-1)^k \frac{\Gamma(N+1)}{\Gamma(N/2+k+1)\Gamma(N/2-k+1)} f(t-kh) \tag{5.18}$$

Definition 5.2 Let N be a positive odd integer. We define a two-sided difference by:

$$\Delta_o^N f(t) = (-1)^{(N+1)/2} \sum_{k=-(N-1)/2}^{(N+1)/2} \times (-1)^k \frac{\Gamma(N+1)}{\Gamma((N+1)/2-k+1)\Gamma((N-1)/2+k+1)} f(t-kh+h/2) \tag{5.19}$$

with these definitions we are able to define the corresponding derivatives.

Definition 5.3 Let N be a positive even integer. We define an even order two-sided derivative by:

$$\begin{aligned} D_e^N f(t) &= \lim_{h\to 0}\frac{\Delta_e^N f(t)}{h^N} \\ &= \lim_{h\to 0}\frac{(-1)^{N/2}}{h^N}\sum_{k=-N/2}^{N/2}(-1)^k\frac{\Gamma(N+1)}{\Gamma(N/2+k+1)\Gamma(N/2-k+1)}f(t-kh) \end{aligned} \tag{5.20}$$

Definition 5.4 Let N be a positive odd integer. We define an odd order two-sided derivative by:

$$D_o^N f(t) = \lim_{h\to 0}\frac{\Delta_o^N f(t)}{h^N} = \lim_{h\to 0}\frac{(-1)^{(N+1)/2}}{h^N}$$
$$\sum_{k=-(N-1)/2}^{(N+1)/2}(-1)^k\frac{\Gamma(N+1)}{\Gamma((N+1)/2-k+1)\Gamma((N-1)/2+k+1)}f(t-kh+h/2) \tag{5.21}$$

Both derivatives (5.20) and (5.21) coincide with the usual derivative Nth order derivative.

5.3 Integral Representations for the Integer Order Two-Sided Differences

The result stated in (5.20) can be interpreted in terms of the residue theorem leading to an integral representation for the difference. Assume that $f(z)$ is analytic inside and on a closed integration path that includes the points $t = z - kh$, $h \in C$, with $k = -N/2, -N/2 + 1, \ldots, -1, 0, 1, \ldots, N/2 - 1, N/2$. Then

$$\Delta_e^N f(z) = \frac{(-1)^{N/2}N!}{2\pi ih}\int_{C_c} f(z+w)\frac{\Gamma\left(\frac{-w}{h}+1\right)\quad\Gamma\left(\frac{w}{h}\right)}{\Gamma\left(\frac{-w}{h}+\frac{N}{2}+1\right)\Gamma\left(\frac{w}{h}+\frac{N}{2}+1\right)}dw \tag{5.22}$$

To prove this result, remark that Eq. 5.20 can be considered as $1/2\pi i\sum$ residues in the computation of the integral of a function with poles at $t = z - kh$ (Fig. 5.1).

We can make a translation and consider poles at kh. As it can be seen by direct verification, we have

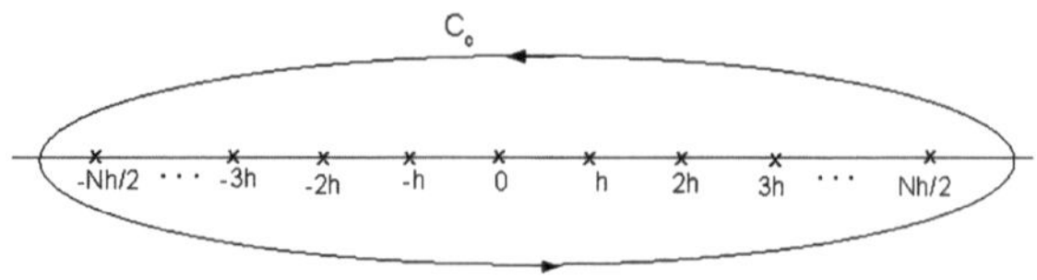

Fig. 5.1 Integration path and poles for the integral representation of integer even order differences

$$\sum_{k=-N/2}^{N/2}\frac{N!(-1)^{N/2-k}}{(N/2+k)!(N/2-k)!}f(t-kh)$$
$$=\frac{N!}{2\pi ih}\int_{C_c}\frac{f(z+w)}{\prod_{k=0}^{N/2}\left(\frac{w}{h}-k\right)\prod_{k=1}^{N/2}\left(\frac{w}{h}+k\right)}\mathrm{d}w \tag{5.23}$$

Introducing the Pochhammer symbol, we can rewrite the above formula as:

$$\Delta_e^N f(z)=\frac{(-1)^{N/2}N!}{2\pi ih}\int_{C_c}\frac{f(z+w)\left(\frac{-w}{h}\right)}{\left(\frac{w}{h}\right)_{N/2+1}\left(\frac{-w}{h}\right)_{N/2+1}}\mathrm{d}w \tag{5.24}$$

Attending to the relation between the Pochhammer symbol and the Gamma function:

$$\Gamma(z+n)=(z)_n\Gamma(z) \tag{5.25}$$

we can write (5.22).

It is easy to test the coherency of (5.22) relatively to (5.20), by noting that the Gamma function $\Gamma(z)$ has poles at the negative integers ($z=-n,\ n\in Z^{+}$). The corresponding residues are equal to $(-1)^n/n!$. Both the Gamma functions have infinite poles, but outside the integration path they cancel out and the integrand is analytic.

Similarly to the above development, we have[2]:

$$\Delta_o^N f(z)=\frac{(-1)^{(N+1)/2}N!}{2\pi ih}\int_{C_c}f(z+w)\frac{\Gamma\left(-\frac{w}{h}+\frac{1}{2}\right)\quad\Gamma\left(\frac{w}{h}+\frac{1}{2}\right)}{\Gamma\left(\frac{-w}{h}+\frac{N}{2}+1\right)\Gamma\left(\frac{w}{h}+\frac{N}{2}+1\right)}\mathrm{d}w \tag{5.26}$$

To prove this, we proceed as above. By direct verification, we have

$$\sum_{k=-N/2}^{N/2}(-1)^{N/2-k}\binom{N}{N/2-k}f(z-kh)$$
$$=\frac{N!}{2\pi ih}\int_{C_c}\frac{f(z+w)}{\prod_{k=0}^{(N-1)/2}\left(\frac{w}{h}-k-\frac{1}{2}\right)\prod_{k=1}^{(N-1)/2}\left(\frac{w}{h}+k+\frac{1}{2}\right)}\mathrm{d}w \tag{5.27}$$

and

[2] Figure 5.2 shows the integration path and corresponding poles.

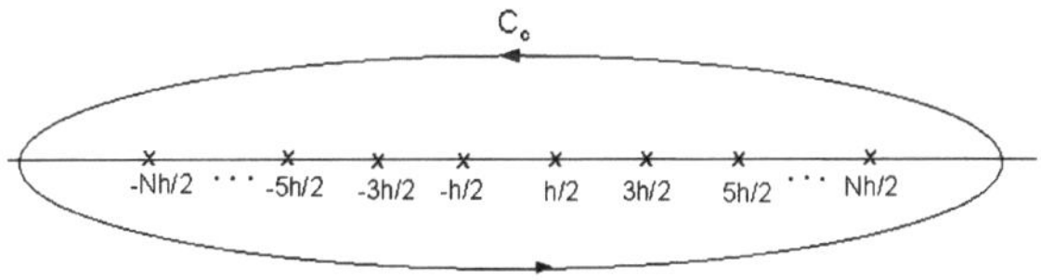

Fig. 5.2 Integration path and poles for the integral representation of integer odd order differences

$$\Delta_o^N f(z) = \frac{(-1)^{(N+1)/2} N!}{2\pi i h} \int_{C_c} \frac{f(z+w)}{\left(\frac{w}{h}+\frac{1}{2}\right)_{(N+1)/2}\left(-\frac{w}{h}+\frac{1}{2}\right)_{(N+1)/2}} \mathrm{d}w \tag{5.28}$$

that leads immediately to (5.26)

5.4 Fractional Central Differences

We are going to consider two types of fractional central differences. Let $\alpha > -1$, $h \in R^+$ and $f(t)$ a Fig. 5.2 complex variable function.

Definition 5.5 We define a type 1 fractional difference by:

$$\Delta_{c_1}^{\alpha} f(t) = \sum_{-\infty}^{+\infty} \frac{(-1)^k \Gamma(\alpha+1)}{\Gamma(\alpha/2-k+1)\Gamma(\alpha/2+k+1)} f(t-kh) \tag{5.29}$$

Definition 5.6 We define a type 2 fractional difference by[3]:

$$\Delta_{c_2}^{\alpha} f(t) = \sum_{-\infty}^{+\infty} \frac{(-1)^k \Gamma(\alpha+1) f(t-kh+h/2)}{\Gamma[(\alpha+1)/2-k+1]\Gamma[(\alpha-1)/2+k+1]} \tag{5.30}$$

Remark that we did not insert any power of (-1). Although it may be useful in some problems to keep it, we found better to remove it due to the relation with the Riesz potentials that we will obtain later.

With the following relation [3][4]:

$$\sum_{-\infty}^{+\infty} \frac{1}{\Gamma(a-k+1)\Gamma(b-k+1)\Gamma(c+k+1)\Gamma(d+k+1)}$$
$$= \frac{\Gamma(a+b+c+d+1)}{\Gamma(a+c+1)\Gamma(b+c+1)\Gamma(a+d+1)\Gamma(b+d+1)} \tag{5.31}$$

valid for $a+b+c+d > -1$, it is not very hard to show that:

[3] Here we assume that α is also non zero.

[4] See page 123.

$$\Delta^{\beta}_{c_1}\{\Delta^{\alpha}_{c_1} f(t)\} = \Delta^{\alpha+\beta}_{c_1} f(t) \tag{5.32}$$

and

$$\Delta^{\beta}_{c2}\{\Delta^{\alpha}_{c_2} f(t)\} = -\Delta^{\alpha+\beta}_{c_1} f(t) \tag{5.33}$$

while

$$\Delta^{\beta}_{c_2}\{\Delta^{\alpha}_{c_1} f(t)\} = \Delta^{\alpha+\beta}_{c_2} f(t) \tag{5.34}$$

provided that $\alpha + \beta > -1$. In particular, $\alpha + \beta = 0$, and the relations (5.32) and (5.33) show that when $|\alpha| < 1$ and $|\beta| < 1$ the inverse differences exist and can be obtained by using formulae (5.29) and (5.30). We must remark that the zero order type 1 difference is the identity operator and is obtained from (5.29). The zero order type 2 difference will be considered later.

5.5 Integral Representations for the Fractional Two-Sided Differences

Let us assume that $f(z)$ is analytic in a region of the complex plane that includes the real axis. To obtain the integral representations for the previous differences we follow here the procedure used in Chap. 3. We only have to give interpretations to (5.29) and (5.30) in terms of the residue theorem. For the first case, we must remark that the poles must lie at nh, $n \in Z$. This leads easily to

$$\Delta^{\alpha}_{C_1} f(t) = \frac{\Gamma(\alpha+1)}{2\pi i j h} \int_{C_c} f(z+w) \frac{\Gamma\left(\frac{-w}{h}+1\right)}{\Gamma\left(\frac{-w}{h}+\frac{\alpha}{2}+1\right)} \frac{\Gamma\left(\frac{w}{h}\right)}{\Gamma\left(\frac{w}{h}+\frac{\alpha}{2}+1\right)} dw \tag{5.35}$$

The integrand function has infinite poles at every nh, with $n \in Z$. The integration path must consist of infinite lines above and below the real axis closing at the infinite. The easiest situation is obtained by considering two straight lines near the real axis, one above and the other below (see Fig. 5.3).

Regarding to the second case, the poles are located now at the half integer multiples of h, which leads to

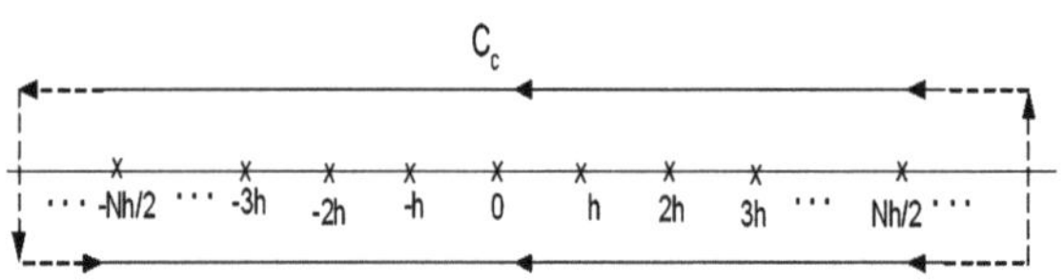

Fig. 5.3 Path and poles for the integral representation of type 1 differences

$$\Delta_{C_2}^{\alpha} f(t) = \frac{\Gamma(\alpha+1)}{2\pi ijh} \int_{C_c} f(z+w) \frac{\Gamma\left(-\frac{w}{h}+\frac{1}{2}\right) \quad \Gamma\left(\frac{w}{h}+\frac{1}{2}\right)}{\Gamma\left(\frac{-w}{h}+\frac{\alpha}{2}+1\right)\Gamma\left(\frac{w}{h}+\frac{\alpha}{2}+1\right)} \mathrm{d}w \tag{5.36}$$

These integral formulations will be used in the following section to obtain the integral formulae for the central derivatives generalising the Cauchy derivative for the two-sided case. We could consider the poles as lying over any straight line as we did with the forward and backward cases in Chap. 3. However, this may not be very important. So we will work over the real axis.

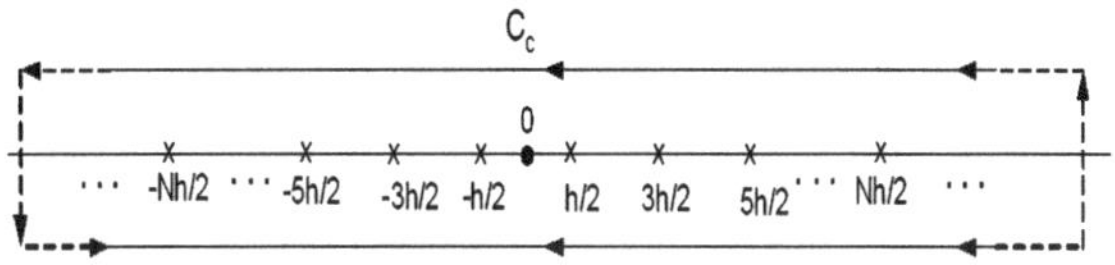

Fig. 5.4 Path and poles for the integral representation of type 2 differences

5.6 The Fractional Two-Sided Derivatives

To obtain fractional central derivatives we proceed as usually [4–8]: divide the fractional differences by h^{α} ($h \in R^{+}$) and let $h \to 0$. For the first case and assuming again that $\alpha > -1$, we obtain:

$$\begin{aligned} D_{c_1}^{\alpha} f(t) &= \lim_{h\to 0} \frac{\Delta_{c_1}^{\alpha} f(t)}{h^{\alpha}} \\ &= \lim_{h\to 0} \frac{\Gamma(\alpha+1)}{h^{\alpha}} \sum_{-\infty}^{+\infty} \frac{(-1)^k}{\Gamma(\alpha/2-k+1)\Gamma(\alpha/2+k+1)} f(t-kh) \end{aligned} \tag{5.37}$$

that we will call type 1 two-sided fractional derivative.

For the second case and assuming also that $\alpha \neq 0$, we obtain the type 2 two-sided fractional derivative given by

$$\begin{aligned} D_{c_2}^{\alpha} f(t) &= \lim_{h\to 0} \frac{\Delta_{c_2}^{\alpha} f(t)}{h^{\alpha}} \\ &= \lim_{h\to 0} \frac{\Gamma(\alpha+1)}{h^{\alpha}} \sum_{-\infty}^{+\infty} \frac{(-1)^k f(t-kh+h/2)}{\Gamma[(\alpha+1)/2-k+1]\Gamma[(\alpha-1)/2+k+1]} \end{aligned} \tag{5.38}$$

Formulae (5.37) and (5.38) generalise the positive integer order central derivatives to the fractional case, although there should be an extra factor $(-1)^{\alpha/2}$ in the first case and $(-1)^{(\alpha+1)/2}$ in the second case that we removed, as referred before.

5.7 Integral Formulae

To obtain the integral formulae for the derivatives we must substitute the integral formulae (5.35) and (5.36) into (5.37) and (5.38) respectively and permute there the limit and integral operations. With this permutation we must compute the limit of two quotients of Gamma functions. As it is well known, the quotient of two gamma functions $\frac{\Gamma(s+a)}{\Gamma(s+b)}$ has the expansion

$$\frac{\Gamma(s+a)}{\Gamma(s+b)} = s^{a-b}\left[1 + \sum_{1}^{N} c_k s^{-k} + O(s^{-N-1})\right] \tag{5.39}$$

as $|s| \to \infty$, uniformly in every sector that excludes the negative real half-axis. When h is very small

$$\frac{\Gamma(w/h+a)}{\Gamma(w/h+b)} \approx (w/h)^{a-b}[1 + h \cdot \varphi(w/h)] \tag{5.40}$$

where φ is regular near the origin. Accordingly to the above statement, the branch cut line used to define a function on the right hand side in (5.40) is the negative real half axis. Similarly, we

$$\frac{\Gamma(-w/h+a)}{\Gamma(-w/h+b)} \approx (w/h)^{a-b}[1 + h \cdot \varphi(-w/h)] \tag{5.41}$$

but now, the branch cut line is the positive real axis. With these results, we obtain generalisations of the Cauchy integral formulation for the type 1 derivative given by

$$D^{\alpha}_{C_1} f(t) = \frac{\Gamma(\alpha+1)}{2\pi i} \int_{C_c} f(z+w) \frac{1}{(w)_l^{\alpha/2+1}(-w)_r^{\alpha/2}} \mathrm{d}w \tag{5.42}$$

while for the type 2 derivative is

$$D^{\alpha}_{C_2} f(t) = \frac{\Gamma(\alpha+1)}{2\pi i} \int_{C_c} f(z+w) \frac{1}{(w)_l^{(\alpha+1)/2}(-w)_r^{(\alpha+1)/2}} \mathrm{d}w \tag{5.43}$$

The subscripts "l" and "r" mean respectively that the power functions have the left and right half real axis as branch cut lines.

Now, we are going to compute the above integrals for the special case of straight line paths. Let us assume that the distance between the horizontal straight lines in Figs. 5.3 and 5.4 is $2\varepsilon(h)$ that decreases to zero with h. In Fig. 5.5 we show the different segments used for the computation of the above integrals. If we assume that the two straight lines are infinitely near, we have for the type 1 derivative:

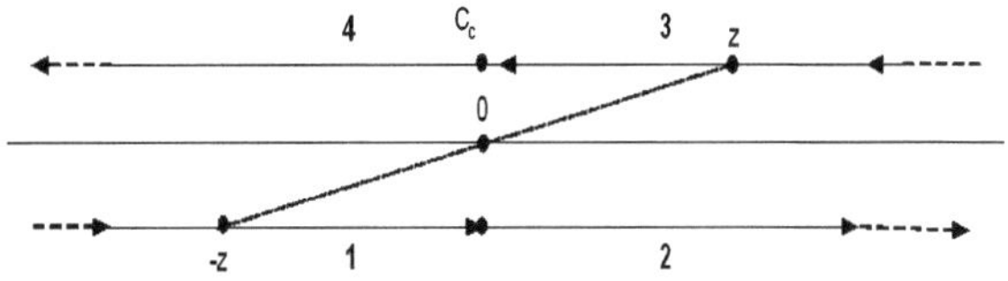

Fig. 5.5 Segments for the computation of the integrals (5.42) and (5.43)

$$\int_1 = -\frac{\Gamma(\alpha+1)}{2\pi i}\int_0^\infty f(z-x)\frac{1}{x^{\alpha+1}e^{-i\alpha\pi/2}e^{-i\pi}}e^{-i\pi}\,dx$$

$$\int_2 = \frac{\Gamma(\alpha+1)}{2\pi i}\int_0^\infty f(z+x)\frac{1}{x^{\alpha+1}e^{i\alpha\pi/2}}\,dx$$

$$\int_3 = -\frac{\Gamma(\alpha+1)e^{-i\alpha\pi/2}}{2\pi i}\int_0^\infty f(z+x)\frac{1}{x^{\alpha+1}e^{-i\alpha\pi/2}}\,dx$$

$$\int_4 = \frac{\Gamma(\alpha+1)}{2\pi i}\int_0^\infty f(z-x)\frac{1}{x^{\alpha+1}e^{i\alpha\pi/2}}e^{i\pi}dx$$

where the integer numbers refer the straight-line segment used in the computation. Joining the four integrals, we obtain:

$$D^\alpha_{C_1}f(t) = -\frac{\Gamma(\alpha+1)\sin(\alpha\pi/2)}{\pi}\int_0^\infty f(z-x)\frac{1}{x^{\alpha+1}}\,dx - \frac{\Gamma(\alpha+1)\sin(\alpha\pi/2)}{\pi}\int_0^\infty f(z+x)\frac{1}{x^{\alpha+1}}\,dx$$

or

$$D^\alpha_{C_1}f(t) = -\frac{\Gamma(\alpha+1)\sin(\alpha\pi/2)}{\pi}\int_{-\infty}^\infty f(z-x)\frac{1}{|x|^{\alpha+1}}\,dx \tag{5.44}$$

As α is not an odd integer and using the reflection formula of the gamma function we obtain

$$D^\alpha_{C_1}f(t) = -\frac{1}{2\Gamma(-\alpha)\cos(\alpha\pi/2)}\int_{-\infty}^\infty f(z-x)\frac{1}{|x|^{\alpha+1}}\,dx \tag{5.45}$$

When $-1 < \alpha < 0$, it is the so called Riesz potential [8], for $0 < \alpha < 1$, it is the corresponding inverse operator.

For the type 2 case, we compute again the integrals corresponding to the four segments to obtain:

$$\int_1 = -\frac{\Gamma(\alpha+1)}{2\pi i}\int_0^\infty f(z-x)\frac{1}{x^{\alpha+1}e^{-i(\alpha+1)\pi/2}}e^{i\pi}dx,$$

$$\int_2 = \frac{\Gamma(\alpha+1)}{2\pi i}\int_0^\infty f(z+x)\frac{1}{x^{\alpha+1}e^{i(\alpha+1)\pi/2}}dx$$

$$\int_3 = -\frac{\Gamma(\alpha+1)e^{-i\alpha\pi/2}}{2\pi i}\int_0^\infty f(z+x)\frac{1}{x^{\alpha+1}e^{-i(\alpha+1)\pi/2}}dx$$

$$\int_4 = \frac{\Gamma(\alpha+1)}{2\pi i}\int_0^\infty f(z-x)\frac{1}{x^{\alpha+1}e^{-i(\alpha+1)\pi/2}}e^{i\pi}dx$$

Joining the four integrals, we obtain:

$$D^\alpha_{c_2}f(t) = \frac{\Gamma(\alpha+1)\sin[(\alpha+1)\pi/2]}{\pi}\int_0^\infty f(z-x)\frac{1}{x^{\alpha+1}}dx - \frac{\Gamma(\alpha+1)\sin[(\alpha+1)\pi/2]}{\pi}\int_0^\infty f(z+x)\frac{1}{x^{\alpha+1}}dx$$

As the last integral can be rewritten as:

$$\int_0^\infty f(z+x)\frac{1}{x^{\alpha+1}}dx = \int_{-\infty}^0 f(z-x)\frac{1}{(-x)^{\alpha+1}}dx$$

we obtain

$$D^\alpha_{c_2}f(t) = -\frac{1}{2\Gamma(-\alpha)\sin(\alpha\pi/2)}\int_{-\infty}^\infty f(z-x)\frac{\mathrm{sgn(x)}}{|x|^{\alpha+1}}dx \tag{5.46}$$

that is the modified Riesz potential [8], when $-1 < \alpha < 0$, when $0 < \alpha < 1$, it is the corresponding inverse operator. Both potentials (5.45) and (5.46) were studied also by Okikiolu [2]. These are essentially convolutions of a given function with two acausal (neither causal nor anti-causal) operators.

Letting $F(\omega)$ be the Fourier transform of $f(t)$ and, as the Fourier transform of $\frac{1}{2\Gamma(-\alpha)\cos(\alpha\pi/2)}|t|^{-\alpha-1}$ is given by $|\omega|^\alpha$ we conclude that:

$$F[D_{c_1}^{\alpha} f(t)] = |\omega|^{\alpha} F(\omega) \tag{5.47}$$

Similarly, as the Fourier transform of $\frac{-\text{sgn}(t)}{(\alpha+1)2\Gamma(-\alpha-1)\cos[(\alpha+1)\pi/2]}|t|^{-\alpha-1}$ is given by $-j|\omega|^{\alpha}\text{sgn}(\omega)$ [2], we conclude that:

$$F[D_{c_2}^{\alpha} f(t)] = -j|\omega|^{\alpha}\text{sgn}(\omega)F(\omega) \tag{5.48}$$

It is interesting to use the type 1 derivative with $\alpha = 2M + 1$ and the type 2 with $\alpha = 2M$. This will be done later.

Relations (5.47) and (5.48) generalise a well known property of the Fourier transform.

5.8 Coherence of the Definitions

5.8.1 Type 1 Derivative

We want to test the coherence of the results by considering functions with Fourier transform. To perform this study, we only have to find the behaviour of the defined derivatives for $f(t) = e^{-j\omega t}$, $t,\omega \in R$. In the following we will consider non integer orders greater than -1. We start by considering the type 1 derivative. From (5.29) we obtain

$$\Delta_{c_1}^{\alpha} e^{j\omega t} = e^{-j\omega t} \sum_{-\infty}^{+\infty} \frac{(-1)^n \Gamma(\alpha+1)}{\Gamma(\alpha/2 - n + 1)\Gamma(\alpha/2 + n + 1)} e^{j\omega nh} \tag{5.49}$$

where we recognize the discrete-time Fourier transform of $R_b(n)$,[5] given by:

$$R_b(n) = \frac{(-1)^n \Gamma(\alpha+1)}{\Gamma(\alpha/2 - n + 1)\Gamma(\alpha/2 + n + 1)} \tag{5.50}$$

As before, this function is the discrete autocorrelation of

$$h_n = \frac{(-\alpha/2)_n}{n!} u_n \tag{5.51}$$

where u_n is the discrete unit step Heaviside function. As the binomial series is convergent over the unit circle excepting the point $z = 1$, the discrete-time Fourier transform of h_n is:

$$H(e^{j\omega}) = FT[h_n] = (1 - e^{-j\omega h})^{\alpha/2} \tag{5.52}$$

and the discrete-time Fourier transform of $R_b(n)$

---

[5] In purely mathematical terms it is a Fourier series with $R_b(n)$ as coefficients.

$$S(e^{j\omega}) = \lim_{z\to e^{j\omega h}} (1 - z^{-1})^{\alpha/2}(1 - z)^{\alpha/2} = (1 - e^{-j\omega h})^{\alpha/2}(1 - e^{j\omega h})^{\alpha/2}$$
$$= \left|e^{j\omega h/2} - e^{-j\omega h/2}\right|^{\alpha} = |2\sin(\omega h/2)|^{\alpha} \quad (5.53)$$

So,

$$|2\sin(\omega h/2)|^{\alpha} = \sum_{-\infty}^{+\infty} \frac{(-1)^n \Gamma(\alpha+1)}{\Gamma(\alpha/2 - n + 1)\Gamma(\alpha/2 + n + 1)} e^{j\omega nh} \quad (5.54)$$

We write, then:

$$\Delta_{c_1}^{\alpha} e^{-j\omega t} = e^{-j\omega t}|2\sin(\omega h/2)|^{\alpha} \quad (5.55)$$

So, there is a linear system with frequency response given by:

$$H_{\Delta 1}(\omega) = |2\ \sin(\omega h/2)|^{\alpha} \quad (5.56)$$

that acts on a signal giving its type 1 central fractional difference. Dividing (5.56) by $h^{\alpha}(h \in R^{+})$ and computing the limit as $h \to 0$, it comes:

$$H_{D1}(\omega) = |\omega|^{\alpha} \quad (5.57)$$

As α is not an even integer:

$$|\omega|^{\alpha} = \lim_{h\to 0} \frac{1}{h^{\alpha}} \sum_{-\infty}^{+\infty} \frac{(-1)^n \Gamma(\alpha+1)}{\Gamma(\alpha/2 - n + 1)\Gamma(\alpha/2 + n + 1)} e^{j\omega nh} \quad (5.58)$$

valid for $\alpha > -1$. The inverse Fourier Transform of $|\omega|^{\alpha}$ is given by Okikiolu [2]:

$$FT^{-1}[|\omega|^{\alpha}] = \frac{1}{2\Gamma(-\alpha)\cos(\alpha\pi/2)} |t|^{-\alpha-1} \quad (5.59)$$

and we obtain the impulse response:

$$h_{D1}(t) = \frac{1}{2\Gamma(-\alpha)\cos(\alpha\pi/2)} |t|^{-\alpha-1} \quad (5.60)$$

leading to

$$D_{C_1}^{\alpha} f(t) = \frac{1}{2\Gamma(-\alpha)\cos(\alpha\pi/2)} \int_{-\infty}^{+\infty} f(\tau)|t-\tau|^{-\alpha-1} d\tau \quad (5.61)$$

that is coincides with (5.45). Relations (5.52) and (5.53) allow us to conclude that the type 1 central derivative is equivalent to the application of the $\alpha/2$ order forward (or backward) derivative twice: one with increasing time and the other with reverse time.

5.8.2 Type 2 Derivative

A similar procedure allows us to obtain

$$\Delta_{c_2}^{\alpha} \mathrm{e}^{-j\omega t} = \mathrm{e}^{-j\omega t} \mathrm{e}^{-j\omega h/2} \sum_{-\infty}^{+\infty} \frac{(-1)^k \Gamma(\alpha+1)}{\Gamma[(\alpha+1)/2-k+1]\Gamma[(\alpha-1)/2+k+1]} \mathrm{e}^{j\omega kh} \tag{5.62}$$

In order to maintain the coherence with the usual definition of discrete-time Fourier transform, we change the summation variable, obtaining

$$\Delta_{c_2}^{\alpha} \mathrm{e}^{-j\omega t} = \mathrm{e}^{-j\omega t} \mathrm{e}^{-j\omega h/2} \sum_{-\infty}^{+\infty} \frac{(-1)^k \Gamma(\alpha+1)}{\Gamma[(\alpha+1)/2+k+1]\Gamma[(\alpha-1)/2-k+1]} \mathrm{e}^{-j\omega kh} \tag{5.63}$$

Now, the coefficients of the above Fourier series are the cross-correlation, $R_{bc}(k)$, between

$$h_n = \frac{(-a)_n}{n!} u_n \tag{5.64}$$

and

$$g_n = \frac{(-b)_n}{n!} u_n \tag{5.65}$$

with $a = (\alpha + 1)/2$ and $b = (\alpha - 1)/2$. Let $S_{bc}(\mathrm{e}^{j\omega})$ be the discrete-time Fourier transform of the cross-correlation, $R_{bc}(k)$:

$$S_{bc}(\mathrm{e}^{j\omega}) = FT[R_{bc}(k)] \tag{5.66}$$

$R_{bc}(k)$ being a correlation, we conclude easily that $S_{bc}(\mathrm{e}^{j\omega})$ is given by:

$$S_{bc}(\mathrm{e}^{j\omega}) = \lim_{z \to \mathrm{e}^{j\omega h}} (1 - z^{-1})^{(\alpha+1)/2} (1 - z)^{(\alpha-1)/2} \tag{5.67}$$

$$= (1 - \mathrm{e}^{-j\omega h})^{(\alpha+1)/2} (1 - \mathrm{e}^{j\omega h})^{(\alpha+1)/2} (1 - \mathrm{e}^{j\omega h})^{-1} \tag{5.68}$$

We write, then:

$$\Delta_{c_2}^{\alpha} \mathrm{e}^{-j\omega t} = \mathrm{e}^{j\omega t} |2 \sin(\omega h/2)|^{\alpha+1} [2j \sin(\omega h/2)]^{-1}$$

So, there is a linear system with frequency response given by:

$$H_{\Delta 2}(\omega) = |2 \sin(\omega h/2)|^{\alpha+1} [2j \sin(\omega h/2)]^{-1} \tag{5.69}$$

that acts on a signal giving its fractional central difference. We can write also

$$|2\sin(\omega h/2)|^{\alpha+1}[2j\sin(\omega h/2)]^{-1}$$
$$= \sum_{-\infty}^{+\infty} \frac{(-1)^k\Gamma(\alpha+1)}{\Gamma[(\alpha+1)/2+k+1]\Gamma[(\alpha-1)/2-k+1]} e^{-j\omega kh} \tag{5.70}$$

Dividing (5.69) by h^α ($h \in R^+$) and computing the limit as $h \to 0$, it gives:

$$H_{D2}(\omega) = -j|\omega|^\alpha \mathrm{sgn}(\omega) \tag{5.71}$$

As

$$j\frac{\mathrm{d}|\omega|^{\alpha+1}}{\mathrm{d}\omega} = j(\alpha+1)|\omega|^\alpha \mathrm{sgn}(\omega)$$

and using a well known property of the Fourier transform we obtain from (5.59):

$$h_{D2}(t) = \frac{-\mathrm{sgn}(t)}{(\alpha+1)2\Gamma(-\alpha-1)\cos[(\alpha+1)\pi/2]}|t|^{-\alpha-1} \tag{5.72}$$

or, using the properties of the gamma function

$$h_{D2}(t) = -\frac{\mathrm{sgn}(t)}{2\Gamma(-\alpha)\sin(\alpha\pi/2)}|t|^{-\alpha-1} \tag{5.73}$$

and as previously:

$$D^\alpha_{C_2} f(t) = -\frac{1}{2\Gamma(-\alpha)\sin(\alpha\pi/2)} \int_{-\infty}^{+\infty} f(\tau)|t-\tau|^{-\alpha-1}\mathrm{sgn}(t-\tau)\mathrm{d}\tau \tag{5.74}$$

Relations (5.64), to (5.68) allow us to conclude that the type 2 central derivative is equivalent to the application of the forward (or backward) derivative twice: one with increasing time and order $(\alpha + 1)/2$, and the other with reverse time and order $(\alpha - 1)/2$.

It is interesting to remark that combining (5.57) with (5.71) as

$$H_D(\omega) = H_{D1}(\omega) + jH_{D2}(\omega) \tag{5.75}$$

We obtain a function that is null for $\omega < 0$. This means that the operator defined by (5.74) is the Hilbert transform of that defined in (5.61) and the corresponding "analytic" derivative is given by the convolution of the function at hand with the operator:

$$h_D(t) = \frac{|t|^{-\alpha-1}}{2\Gamma(-\alpha)\cos(\alpha\pi/2)} - j\frac{|t|^{-\alpha-1}\mathrm{sgn}(t)}{2\Gamma(-\alpha)\sin(\alpha\pi/2)} \tag{5.76}$$

This is formally similar to the Riesz–Feller potential definitions [8].

5.8.3 The Integer Order Cases

Let $\alpha = 2\,N$, $N \in Z^+$, in the type 1 difference. We obtain:

$$\Delta_{c_1}^{2N} f(t) = \sum_{-N}^{+N} \frac{(-1)^k (2N)!}{(N-k)!(N+k)!} f(t-kh) \tag{5.77}$$

that can be written as

$$\Delta_{c_1}^{2N} f(t) = \sum_{-N}^{+N} (-1)^k \binom{2N}{N-K} f(t-kh) \tag{5.78}$$

A close look into (5.78) shows that aside a factor $(-1)^N$ it is the current $2N$ order central difference, as already known. With $N = 0$, we obtain $f(t)$. Similarly, if α is odd ($\alpha = 2N + 1$), the type 2 difference is equal to current central difference, aside the factor $(-1)^{N+1}$. In fact, we have:

$$\Delta_{c_2}^{2N+1} f(t) = \sum_{-N}^{N+1} \frac{(-1)^k (2N+1)! f(t-kh+h/2)}{(N+1-k)!(N+k)!} \tag{5.79}$$

and

$$\Delta_{c_2}^{2N+1} f(t) = \sum_{-N}^{N+1} (-1)^k \binom{2N+1}{N-k} f(t-kh+h/2) \tag{5.80}$$

In particular, with $N = 0$, we obtain

$$\Delta_{c_2}^{1} f(t) = f(t+h/2) - f(t-h/2).$$

It is interesting to use the central type 1 difference (or, derivative) with $\alpha = 2M + 1$ and the type 2 with $\alpha = 2M$. For the first, $\alpha/2$ is not integer and we can use formulae (5.49) to (5.57). However, they are difficult to manipulate. We found better to use (5.59), but we must avoid the product $\Gamma(-\alpha)\cdot\cos(\alpha\pi/2)$, because the first factor is ∞ and the second is zero. We solve the problem by noting that

$$\frac{1}{2\Gamma(-\alpha)\cos(\alpha\pi/2)} = -\frac{\Gamma(\alpha+1)\cdot\sin(\alpha\pi)}{2\pi\cos(\alpha\pi/2)} = -\frac{\Gamma(\alpha+1)\sin(\alpha\pi/2)}{\pi}$$

assuming the value $-\frac{(2M+1)!(-1)^M}{\pi}$. We obtain finally [9]

$$FT^{-1}\left[|\omega|^{2M+1}\right] = -\frac{(2M+1)!(-1)^M}{\pi}|t|^{-2M-2} \tag{5.81}$$

and the corresponding impulse response:

$$h_{D1}(t) = -\frac{(2M+1)!(-1)^M}{\pi}|t|^{-2M-2} \tag{5.82}$$

Relatively to the second case, $\alpha = 2M$, we can use formula (5.73), provided that we use the relation:

$$-\frac{1}{2\Gamma(-\alpha)\sin(\alpha\pi/2)} = \frac{\Gamma(\alpha+1)\cdot\sin(\alpha\pi)}{2\pi\sin(\alpha\pi/2)} = \frac{\Gamma(\alpha+1)\cos(\alpha\pi/2)}{\pi}$$

to get a factor $\frac{(2M)!(-1)^M}{\pi}$. We obtain then [9]:

$$FT^{-1}\left[|\omega|^{2M}\operatorname{sgn}(\omega)\right] = \frac{\operatorname{sgn}(t)(2M)!(-1)^M}{\pi}|t|^{-2M-1} \tag{5.83}$$

and

$$h_{D2}(t) = \frac{\operatorname{sgn}(t)(2M)!(-1)^M}{\pi}|t|^{-2M-1} \tag{5.84}$$

As we can see, the formulae (5.82) and (5.84) allow us to generalize the Riesz potentials for positive integer orders. However, they do not have inverse.

It is interesting to study the situation defined by $\alpha = 0$ with the type 2 derivative. From (5.74) and (5.84) and noting that $t = |t|\cdot\operatorname{sgn}(t)$, we obtain:

$$D^0_{C_2} f(t) = \frac{1}{\pi}\int_{-\infty}^{\infty} f(z-x)\frac{1}{x}\,\mathrm{d}x \tag{5.85}$$

that is the Hilbert transform of $f(t)$.

These results allow us to conclude that:

1. Both type 1 (5.37) and type 2 (5.38) derivatives are defined and meaningful for real orders greater than -1.
2. When the order is an even (odd) integer, type 1 (type 2) derivative is aside a sign equal to the common derivative with the same order.
3. For the same orders, these derivatives cannot be expressed by the Riesz potentials (5.61) and (5.74), because the factors before the integrals are zero.

5.8.4 Other Properties of the Central Derivatives

From the relations (5.32), (5.33), and (5.34) we obtain easily:

$$D^{\beta}_{c_1}\{D^{\alpha}_{c_1} f(t)\} = D^{\alpha+\beta}_{c_1} f(t) \tag{5.86}$$

and

$$D^{\beta}_{c_2}\{D^{\alpha}_{c_2} f(t)\} = -D^{\alpha+\beta}_{c2} f(t) \tag{5.87}$$

while

$$D_{c_3}^{\beta}\{D_{c_3}^{\alpha}f(t)\} = D_{c_3}^{\alpha+\beta}f(t) \tag{5.88}$$

again with $\alpha + \beta > -1$. We conclude:

- If $|\alpha| < 1$ and $|\beta| < 1$ the fractional derivative has always an inverse.
- We can generate the Hilbert transform of a given function with derivations of different types and symmetric orders.

5.9 On the Existence of a Inverse Riesz Potential

In current literature [8] the Riesz potentials are only defined for negative orders verifying $-1 < \alpha < 0$. However, our formulation is valid for every $\alpha > -1$. This means that we can define those potentials even for positive orders. However, we cannot guaranty that there is always an inverse for a given potential. The theory presented in Sect. 5.2 allows us to state that:

- The inverse of a given potential, when existing, is of the same type: the inverse of the type k ($k = 1,2$) potential is a type k potential.
- The inverse of a given potential exists iff its order α verifies $|\alpha| < 1$.
- The order of the inverse of an α order potential is a $-\alpha$ order potential.
- The inverse can be computed both by (5.37) [respectively (5.38)] and by (5.61) [respectively (5.74)].

This is in contradiction with the results stated in Samko et al. [8] about this subject and will have implications in the solution of differential equations involving two-sided derivatives.

5.9.1 Some Computational Issues

In practical applications, we may need to compute a two-sided derivative of a function for which a closed form is not available and we are obliged to truncate the summation or the integral. This leads to an error. We can obtain a bound for such error, by considering a bounded function—$|f(t)| < M$—known inside an interval that we will assume to be symmetric relatively to the origin, $[-L, L]$, only by simplicity. We are going to consider the type 1 case. The other is similar. From (5.61), we conclude that the error is bounded:

$$E < \frac{M}{|\Gamma(-\alpha)\cos(\alpha\pi/2)|}\int_L^{\infty}\frac{1}{x^{\alpha+1}}\mathrm{d}x = \frac{ML^{-\alpha}}{|\Gamma(-\alpha)\cos(\alpha\pi/2)|} = \frac{|\Gamma(\alpha+1)|}{\pi}ML^{-\alpha} \tag{5.89}$$

This result is similar to the one stated by Podlubny [10] in connection with the called there "short-memory" principle. A similar result can be obtained for the summation in (5.37). However, here we have an error bound that is function of h. From the properties of the gamma functions, we obtain easily:

$$\frac{(-1)^k}{\Gamma(\alpha/2-k+1)} = -\frac{\sin(\alpha\pi/2)}{\pi}\Gamma(-\alpha/2+k) \tag{5.90}$$

$$\frac{(-1)^k}{\Gamma(\alpha/2-k+1)\Gamma(\alpha/2+k+1)} = -\frac{\sin(\alpha\pi/2)}{\pi}\frac{\Gamma(-\alpha/2+|k|)}{\Gamma(\alpha/2+|k|+1)} \tag{5.91}$$

When $|k|<1$ is high enough, we can use (5.39) again, to obtain

$$\left|\frac{(-1)^k}{\Gamma(\alpha/2-k+1)\Gamma(\alpha/2+k+1)}\right| \sim \frac{1}{\pi}|k|^{-\alpha-1} \tag{5.92}$$

This leads to an error:

$$E(h) \sim \frac{|\Gamma(\alpha+1)|}{\pi}\sum_{L+1}^{+\infty}|k/h|^{-\alpha-1}h \tag{5.93}$$

and leads to (5.89) again.

5.10 Conclusions

We introduced a general framework for defining the fractional central differences and consider two cases that are generalisations of the usual central differences. These new differences led to central derivatives similar to the usual Grünwald–Letnikov ones. For those differences, we presented integral representations from where we obtained the derivative integrals, similar to Cauchy, by using the properties of the Gamma function. The computation of those integrals led to generalisations of the Riesz potentials. The most interesting feature lies in the summation formulae for the Riesz potentials. To test the coherence of the proposed definitions we applied them to the complex exponential. The results show that they are suitable for functions with Fourier transform. Some properties of these derivatives were presented.

References

1. Henrici P (1991) Applied and computational complex analysis, vol 2. Wiley, New York, pp 389–391
2. Okikiolu GO (1966) Fourier transforms of the operator H_α. Proc Camb Philos Soc 62:73–78

3. Andrews GE, Askey R, Roy R (1999) Special functions. Cambridge University Press, Cambridge
4. Ortigueira MD (2004) From differences to differintegrations. Fract Calc Appl Anal 7(4):459–471
5. Ortigueira MD (2006) A coherent approach to non integer order derivatives. Signal Process Special Sect Fract Calc Appl Signals Syst 86(10):2505–2515
6. Ortigueira MD, Serralheiro AJ (2006) A new least-squares approach to differintegration modelling. Signal Process Special Sect Fract Calc Appl Signals Syst 86(10):2582–2591
7. Ortigueira MD (2006) Riesz potentials and inverses via centred derivatives. Int J Math Math Sci 2006:1–12. Article ID 48391
8. Samko SG, Kilbas AA, Marichev OI (1987) Fractional integrals and derivatives—theory and applications. Gordon and Breach Science Publishers, New York
9. Kilbas AA, Srivastava HM, Trujillo JJ (2006) Theory and applications of fractional differential equations. Elsevier, Amsterdam
10. Podlubny I (1999) Fractional differential equations: an introduction to fractional derivatives, fractional differential equations, to methods of their solution and some of their applications. Academic Press, San Diego

Chapter 6
The Fractional Quantum Derivative and the Fractional Linear Scale Invariant Systems

6.1 Introduction

The normal way of introducing the notion of derivative is by means of the limit of an incremental ratio that can assume three forms, depending the used translations as we saw in Chaps. 2 and 5. On the other hand, in those derivatives the limit operation is done over a set of points uniformly spaced: a linear scale was used. Here we present an alternative derivative, that is valid only for $t > 0$ or $t < 0$ and uses an exponential scale. We are going to introduce the so-called Quantum Derivative [1, 2]. We proceed as in Chap. 2. Let. Δ_q be the following incremental ratio:

$$\Delta_q f(t) = \frac{f(t) - f(qt)}{(1-q)t} \tag{6.1}$$

where q is a positive real number less than 1 and $f(t)$ is assumed to be a causal type signal. The corresponding derivative is obtained by computing the limit as q goes to 1 (to be more precise, we should state $q \to 1^-$)

$$D_q f(t) = \lim_{q\to 1} \frac{f(t) - f(qt)}{(1-q)t} \tag{6.2}$$

This derivative uses values of the variable below t. We can introduce another one that uses values above t. It is defined by

$$D_{q^{-1}} f(t) = \lim_{q\to 1} \frac{f(q^{-1}t) - f(t)}{(q^{-1}-1)t} \tag{6.3}$$

We will generalize these derivatives, first for integer orders, and afterwards for real ones as we did before. We will present the two formulations that come naturally from (6.2) and (6.3) and using values below and above the independent

M. D. Ortigueira, *Fractional Calculus for Scientists and Engineers*, Lecture Notes in Electrical Engineering, 84, DOI: 10.1007/978-94-007-0747-4_6,

variable. We can define also "two-sided" derivatives as we did in Chap. 5, but we will not do it here.

From the Mellin transform of both derivatives we will obtain two integral formulae similar to the Liouville derivatives presented in Chap. 2. Although we will not study here the properties of these derivatives, it may be advanced that they can be used in scale variation problems and to deal with systems defined by Euler–Cauchy type differential equations as we will see later. For now, we will present the steps leading to the fractional quantum derivative and its relation with the Mellin transform (MT).

6.2 The Summation Formulations

6.2.1 The "Below t" Case

We begin by generalizing formula (6.1) for any positive integer order. The formula can be obtained by its repeated application, but we prefer to work in the context of the Mellin Transform due to its simplicity. Let us introduce the Mellin transform by

$$H(s) = \int_0^\infty h(u)u^{-s-1}\mathrm{d}u \tag{6.4}$$

with $s \in C$. This transform is slightly different from the one presented by Bertrand et al. [3] and in current literature, but this one is more convenient since leads to results similar to those obtained with the Laplace transform in the study of shift invariant systems.

Consider that our domain is R^+. We introduce the multiplicative convolution defined by

$$f(t) \vee g(t) = \int_0^\infty f(t/v)g(v)\frac{\mathrm{d}v}{v} \tag{6.5}$$

It is easy to see that the neutral element of this convolution is $g(t) = \delta(t-1)$. With this, we can show that

$$\Delta_q f(t) = \left[\frac{\delta(t-1) - \delta(t-q^{-1})}{(1-q)}\right] \vee [t^{-1}f(t)] \tag{6.6}$$

As it is known, the Mellin Transform of the *multiplicative convolution* is equal to the product of the transforms of both functions. So we obtain:

$$\mathbf{M}[\Delta_q f(t)] = \frac{1-q^{s+1}}{1-q}F(s+1) \tag{6.7}$$

The repeated application of the operator (6.7) leads to:

$$\mathbf{M}[\Delta_q^N f(t)] = \prod_{i=1}^{N} \frac{1-q^{s+i}}{1-q} F(s+N) \tag{6.8}$$

We are going to manipulate the first factor and use the q-binomial formula [1]

$$[1-q^{s+1}]_q^N = \prod_{i=0}^{N-1} (1-q^{1+s}q^i)$$

We have first

$$\prod_{i=1}^{N} \frac{1-q^{s+i}}{1-q} = \frac{\Pi_{i=1}^{N}(1-q^s q^i)}{(1-q)^N} = \frac{\Pi_{i=0}^{N-1}(1-q^{1+s}q^i)}{(1-q)^N} = \frac{[1-q^{s+1}]_q^N}{(1-q)^N}$$

The Gauss binomial formula

$$[a+b]_q^N = \sum_{j=0}^{N} \begin{bmatrix} N \\ j \end{bmatrix}_q (-1)^j q^{j(j-1)/2} b^j a^{N-j}$$

allows us to obtain a different way of expressing the formula on the right. Introducing the q-binomial coefficients

$$\begin{bmatrix} \alpha \\ i \end{bmatrix}_q = \frac{[\alpha]_q!}{[j]_q!(\alpha-i)_q!} \tag{6.9}$$

with $[\alpha]_q$ given by

$$[\alpha]_q = \frac{1-q^\alpha}{1-q} \tag{6.10}$$

the expression on the right can be written as [1, 2]

$$\frac{[1-q^{s+1}]_q^N}{(1-q)^N} = \frac{\sum_{j=0}^{N} \left[\begin{smallmatrix} N \\ j \end{smallmatrix}\right]_q (-1)^j q^{j(j+1)/2} q^{js}}{(1-q)^N} \tag{6.11}$$

that inserted into (6.8) gives:

$$\mathbf{M}\left[\Delta_q^N f(t)\right] = \frac{\sum_{j=0}^{N} \begin{bmatrix} N \\ j \end{bmatrix}_q (-1)^j q^{j(j+1)/2} q^{js}}{(1-q)^N} F(s+N) \tag{6.12}$$

From the properties of the Mellin transform [4]

$$\mathbf{M}^{-1}[q^{js}F(s+N)] = q^{-jN} t^{-N} f(q^j t) \tag{6.13}$$

We conclude that:

$$\Delta_q^N f(t) = t^{-N} \frac{\sum_{j=0}^{N} \begin{bmatrix} N \\ j \end{bmatrix}_q (-1)^j q^{j(j+1)/2} q^{-jN} f(q^j t)}{(1-q)^N} \tag{6.14}$$

To obtain the corresponding derivatives we only have to perform the limit computation [1, 2, 5]

$$D_q^N f(t) = t^{-N} \lim_{q\to 1} \frac{\sum_{j=0}^{N} \begin{bmatrix} N \\ j \end{bmatrix}_q (-1)^j q^{j(j+1)/2} q^{-jN} f(q^j t)}{(1-q)^N} \tag{6.15}$$

To test the behaviour of the above formula, let us compute the second derivative of the function $f(t) = t^3 u(t)$, where $u(t)$ is the Heaviside unit step. We have:

$$D_q^2 f(t) = t \lim_{q\to 1} \frac{\sum_{j=0}^{2} \begin{bmatrix} 2 \\ j \end{bmatrix}_q (-1)^j q^{j(j+1)/2} q^j}{(1-q)^2}$$

and, from (6.11),

$$\begin{aligned} D_q^2 f(t) &= t \lim_{q\to 1} \frac{\Pi_{i=0}^{1}(1-q^2 q^i)}{(1-q)^2} = t \lim_{q\to 1} \frac{(1-q^2)(1-q^3)}{(1-q)^2} \\ &= t \lim_{q\to 1}(1+q)(1+q+q^2) = 6t \end{aligned}$$

as expected.

From (6.8) and (6.12), we conclude that:

$$\mathbf{M}\left[t^{-N} \frac{\sum_{j=0}^{N} \begin{bmatrix} N \\ j \end{bmatrix}_q (-1)^j q^{j(j+1)/2} q^{-jN} f(q^j t)}{(1-q)^N} \right] = \prod_{i=1}^{N} \frac{1-q^{s+i}}{1-q} F(s+N) \tag{6.16}$$

and, performing the limit computation in the right hand side,

$$\mathbf{M}\left[t^{-N} \lim_{q\to 1} \frac{\sum_{j=0\,q}^{N} \begin{bmatrix} N \\ j \end{bmatrix}_q (-1)^j q^{j(j+1)/2} q^{-jN} f(q^j t)}{(1-q)^N} \right] = (1+s)_N F(s+N) \tag{6.17}$$

where we represented by $(a)_n = a(a+1)\cdots(a+N-1)$ the Pochhammer symbol. From well known properties of the Gamma function, we can write

$$\mathbf{M}\left[t^{-N} \lim_{q\to 1} \frac{\sum_{j=0\,q}^{N} \begin{bmatrix} N \\ j \end{bmatrix} (-1)^j q^{j(j+1)/2} q^{-jN} f(q^j t)}{(1-q)^N} \right] = \frac{\Gamma(1+s+N)}{\Gamma(1+s)} F(s+N) \tag{6.18}$$

$$= (-1)^N \frac{\Gamma(-s)}{\Gamma(-s-N)} F(s+N) \tag{6.19}$$

The right hand side in (6.18) or (6.19) is the well known Mellin transform of the Nth order derivative. The left side is a new way of expressing such derivative. This expression suggests that we may work with the "derivative"$t^n D_q^N f(t)$, also called *scale derivative*.

We are going to generalise the previous results for the case of a real order, α. So, let us return to (6.11) and substitute α for N in the left hand expression. In the numerator we obtain the fractional q-binomial $[1-q^{s+1}]_q^\alpha$. The generalised Gauss binomial formula [1]

$$[1+a]_q^\alpha = \sum_{j=0}^{\infty} \begin{bmatrix} \alpha \\ j \end{bmatrix}_q (-1)^j q^{j(j-1)/2} a^j$$

allows us to write:

$$\left[1-q^{s+1}\right]_q^\alpha = \sum_{j=0}^{\infty} \begin{bmatrix} \alpha \\ j \end{bmatrix}_q (-1)^j q^{j(j+1)/2} q^{js} \tag{6.20}$$

From the properties of the Mellin transform

$$\mathbf{M}^{-1}[q^{js} F(s+\alpha)] = q^{-j\alpha} t^{-\alpha} f(q^j t) \tag{6.21}$$

This leads us to a Grunwald–Letnikov like fractional quantum derivative:

$$D_q^\alpha f(t) = t^{-\alpha} \lim_{q \to 1} \frac{\sum_{j=0}^{\infty} \begin{bmatrix} \alpha \\ j \end{bmatrix}_q (-1)^j q^{j(j+1)/2} q^{-j\alpha} f(q^j t)}{(1-q)^\alpha} \tag{6.22}$$

that is similar to the formulation proposed by Al-Salam [6]. In (6.22) the q-binomial coefficients are given by

$$\begin{bmatrix} \alpha \\ j \end{bmatrix}_q \frac{[1-q^\alpha]_q^j}{[j]_q} \tag{6.23}$$

Let us introduce the q-gamma function by

$$\Gamma_q(t) \frac{[1-q]_q^\infty}{[1-q^t]_q^\infty (1-q)^{t-1}} \tag{6.24}$$

where $\mathrm{Re}(t) > 0$. With this function,

$$\Gamma_q(n+1) = \frac{[1-q]_q^n}{(1-q)^n} = \prod_{i=1}^{n} \frac{1-q^i}{1-q} = [n]_q! \tag{6.25}$$

the binomial coefficients can be written as:

$$\begin{bmatrix} \alpha \\ j \end{bmatrix}_q = \frac{\Gamma_q(\alpha+1)}{\Gamma_q(\alpha-j+1)\Gamma_q(j+1)} \tag{6.26}$$

On the other hand, the fractional q-binomial in (6.20) is given by

$$\left[1-q^{s+1}\right]_q^{\alpha} = \frac{\left[1-q^{s+1}\right]_q^{\infty}}{\left[1-q^{s+\alpha+1}\right]_q^{\infty}} \tag{6.27}$$

With (6.24), we can write

$$\left[1-q^{s+1}\right]_q^{\alpha} = \frac{\Gamma_q(1+s+\alpha)}{\Gamma_q(1+s)} \cdot (1-q)^{\alpha} \tag{6.28}$$

valid for $\mathrm{Re}(s) > -\min(0,\alpha) - 1$. As the limit of $\Gamma_q(t)$ when $q \to 1$ is $\Gamma(t)$, it is a simple task to obtain:

$$\mathbf{M}\left[D_q^{\alpha} f(t)\right] = \frac{\Gamma(1+s+\alpha)}{\Gamma(1+s)} F(s+\alpha) \tag{6.29}$$

for $\mathrm{Re}(s) > -\min(0,\alpha) - 1$. This relation is the fractional generalisation of the integer order property [3] and allows us to obtain an integral representation of the fractional quantum derivative. We will return to this subject later.

To maintain the coherence we will consider (6.18) as the correct solution in the integer order case, for the "below t" situation.

6.2.2 The "Above t" Case

We are going to study the derivative using a grid of values above t. We proceed as in the last section. Let $\Delta_{q^{-1}}$ be the following incremental ratio:

$$\Delta_{q^{-1}} f(t) = \frac{f(q^{-1}t) - f(t)}{(q^{-1}-1)t} \tag{6.30}$$

With the convolution (6.5), we can show that

$$\Delta_{q^{-1}} f(t) = \left[\frac{\delta(t-q) - \delta(t-1)}{(q^{-1}-1)}\right] v\left[t^{-1} f(t)\right] \tag{6.31}$$

Using the Mellin Transform we obtain:

$$\mathbf{M}\left[\Delta_{q^{-1}} f(t)\right] = \frac{q^{-(s+1)} - 1}{q^{-1}-1} F(s+1) \tag{6.32}$$

The repeated application of the operator (6.32) leads to:

$$\mathbf{M}\left[\Delta_{q^{-1}}^{N} f(t)\right] = \prod_{i=1}^{N} \frac{q^{-(s+i)}}{q^{-1}-1} F(s+N) \tag{6.33}$$

We are going to transform the first factor

$$\prod_{i=1}^{N} \frac{q^{-(s+i)}-1}{q^{-1}-1} = \frac{\Pi_{i=0}^{N-1}\left(1-q^{-s-N}q^{i}\right)}{\left(1-q^{-1}\right)^{N}} = \frac{\left[1-q^{-s-N}\right]_{q}^{N}}{\left(1-q^{-1}\right)^{N}} \tag{6.34}$$

and use the q-binomial formula leading to

$$\mathbf{M}\left[\Delta_{q^{-1}}^{N} f(t)\right] = \frac{\sum_{j=0}^{N} \begin{bmatrix} N \\ j \end{bmatrix}_q (-1)^j q^{j(j-1)/2} q^{-j(s+N)}}{\left(1-q^{-1}\right)^{N}} F(s+N) \tag{6.35}$$

and with (6.16),

$$\Delta_{q^{-1}}^{N} f(t) = t^{-N} \frac{\sum_{j=0}^{N} \begin{bmatrix} N \\ j \end{bmatrix}_q (-1)^j q^{j(j-1)/2} f\left(q^{-j}t\right)}{\left(1-q^{-1}\right)^{N}} \tag{6.36}$$

allowing us to obtain to the derivative:

$$D_{q^{-1}}^{N} f(t) = t^{-N} \lim_{q\to 1} \frac{\sum_{j=0}^{N} \begin{bmatrix} N \\ j \end{bmatrix}_q (-1)^j q^{(j-1)/2} f\left(q^{-j}t\right)}{\left(1-q^{-1}\right)^{N}} \tag{6.37}$$

With the left hand side in (6.35) we conclude that:

$$\mathbf{M}\left[D_{q^{-1}}^{N} f(t)\right] = (1+s)_N F(s+N) \tag{6.38}$$

that coincides with (6.17) as expected.

To generalize the above results for any order, we substitute α for N in the above expressions. We have from (6.35):

$$\mathbf{M}\left[\Delta_{q^{-1}}^{\alpha} f(t)\right] = \frac{\sum_{j=0}^{\infty} \begin{bmatrix} \alpha \\ j \end{bmatrix}_q (-1)^j q^{j(j-1)/2} q^{-j(s+\alpha)}}{\left(1-q^{-1}\right)^{\alpha}} F(s+\alpha) \tag{6.39}$$

and then

$$D_{q^{-1}}^{\alpha} f(t) = t^{-\alpha} \lim_{q\to 1} \frac{\sum_{j=0}^{\infty} \begin{bmatrix} \alpha \\ j \end{bmatrix}_q (-1)^j q^{j(j-1)/2} f\left(q^{-j}t\right)}{\left(1-q^{-1}\right)^{\alpha}} \tag{6.40}$$

Using the q-binomial theorem, we have:

$$\sum_{j=0}^{\infty} \begin{bmatrix} \alpha \\ j \end{bmatrix}_q (-1)^j q^{j(j-1)/2} q^{-j(s+\alpha)} = \left[1-q^{-s-\alpha}\right]_q^{\alpha}$$

and

$$[1-q^{-s-\alpha}]_q^{\alpha} = \frac{[1-q^{-s-\alpha}]_q^{\infty}}{[1-q^{-s}]_q^{\infty}} \tag{6.41}$$

So, with (6.24)

$$[1-q^{-s-\alpha}]_q^{\alpha} = \frac{\Gamma_q(-s)}{\Gamma_q(-s-\alpha)} \cdot (1-q)^{\alpha}$$

and finally

$$\mathbf{M}\left[D_{q^{-1}}^{\alpha} f(t)\right] = (-1)^{\alpha} \cdot \frac{\Gamma(-s)}{\Gamma(-s-\alpha)} F(s+\alpha) \tag{6.42}$$

valid for $\mathrm{Re}(s) < -\max(0, \alpha)$ and in agreement with (6.38) and (6.19).

6.3 Integral Formulations

The two Mellin transforms in (6.29) and (6.42) lead to different integral representation of fractional derivatives by computing the corresponding inverse functions. To do it, we will use well known results of the Beta function. To start, we are going to obtain the inverse $h_b(t)$ of $\frac{\Gamma(-s)}{\Gamma(-s-\alpha)}$.

As known The Euler Beta function is defined for $\mathrm{Re}(p) > 0$ and $\mathrm{Re}(q) > 0$ by

$$B(p,q) = \int_0^1 \tau^{p-1}(1-\tau)^{q-1} \mathrm{d}\tau \tag{6.43}$$

and it can be shown that [7]

$$B(p,q) = \frac{\Gamma(p)\Gamma(q)}{\Gamma(p+q)} \tag{6.44}$$

This allows us to write:

$$\frac{\Gamma(-s)\Gamma(-\alpha)}{\Gamma(-s-\alpha)} = \int_0^1 \tau^{-s-1}(1-\tau)^{-\alpha-1} \mathrm{d}\tau \tag{6.45}$$

Provided that $\mathrm{Re}(s) < 0$ and $\mathrm{Re}(\alpha) < 0$. This gives immediately

$$h_b(t) = \frac{(-1)^{\alpha}}{\Gamma(-\alpha)} (1-t)^{-\alpha-1} u(1-t) \tag{6.46}$$

A similar procedure to obtain the inverse $h_a(t)$ of $\frac{\Gamma(1+s+\alpha)}{\Gamma(1+s)}$ gives

$$\frac{\Gamma(1+s+\alpha)\Gamma(-\alpha)}{\Gamma(1+s)} = \int_0^1 \tau^{1+s+\alpha}(1-\tau)^{-\alpha-1}d\tau \tag{6.47}$$

With a variable change inside the integral, we obtain:

$$\frac{\Gamma(1+s+\alpha)\Gamma(-\alpha)}{\Gamma(1+s)} = \int_1^\infty \tau^{-s}(1-\tau)^{-\alpha-1}d\tau \tag{6.48}$$

and

$$h_a(t) = \frac{1}{\Gamma(-\alpha)}(t-1)^{-\alpha-1}u(t-1) \tag{6.49}$$

To compute in integral formulations of the derivatives corresponding to (6.29) and (6.42) we remark that the inverse Mellin transform of $F(s+\alpha)$ is given by:

$$\mathbf{M}^{-1}[F(s+\alpha)] = t^{-\alpha}f(t) \tag{6.50}$$

and use the convolution (6.5). With (6.46) and (6.49) we obtain the following integral formulations, valid for $\mathrm{Re}(\alpha) < 0$.

$$D_b^\alpha f(t) = -\frac{t^{-\alpha}}{\Gamma(-\alpha)}\int_0^1 f(t/\tau)(1-\tau^{-1})^{-\alpha-1}d\tau \tag{6.51}$$

and

$$D_a^\alpha f(t) = \frac{t^{-\alpha}}{\Gamma(-\alpha)}\int_1^\infty f(t/\tau)(\tau^{-1}-1)^{-\alpha-1}d\tau \tag{6.52}$$

Attending to the fact that the convolution (6.5) is commutative, we can obtain another set of integral formulations for the derivatives. In fact, from (6.46) and (6.49), we obtain:

$$D_a^\alpha f(t) = -\frac{1}{\Gamma(-\alpha)}\int_0^t (t/\tau-1)^{-\alpha-1}\tau^{-\alpha}f(\tau)d\tau/t$$

and

$$D_a^\alpha f(t) = \frac{1}{\Gamma(-\alpha)}\int_0^t (t-\tau)^{-\alpha-1}f(\tau)d\tau \tag{6.53}$$

that coincides with the Liouville derivative particularized for causal functions. Relatively to the other case, we have:

$$D_b^{\alpha} f(t) = -\frac{1}{\Gamma(-\alpha)} \int_t^{\infty} (t-\tau)^{-\alpha-1} f(\tau) \mathrm{d}\tau \tag{6.54}$$

that is the backward Liouville derivative for causal signals. Although we obtained these results for $\alpha < 0$, they remain valid for other values of α, since $\frac{\Gamma(-s)}{\Gamma(-s-\alpha)}$ and $\frac{\Gamma(1+s+\alpha)}{\Gamma(1+s)}$ are analytical in the regions of convergence and we can fix an integration path independent of α. This can be confirmed by expanding (6.46) and (6.49) and transforming each term of the series.

6.4 On the Fractional Linear Scale Invariant Systems

6.4.1 Introduction

Braccini and Gambardella [8] introduced the concept of "form-invariant" filters. These are systems such that a scaling of the input gives rise to the same scaling of the output. This is important in detection and estimation of signals with unknown size requiring some type of pre-processing: for example edge sharpening in image processing or in radar signals. However in their attempt to define such systems, they did not give any formulation in terms of a differential equation. The Linear Scale Invariant Systems (LScIS) were really introduced by Yazici and Kashyap [9, 10] for analysis and modelling 1/f phenomena and in general the self-similar processes, namely the scale stationary processes. Their approach was based on an integer order Euler–Cauchy differential equation. However, they solved only a particular case corresponding to the all pole case. To insert a fractional behaviour, they proposed the concept of pseudo-impulse response. Here we avoid this procedure by presenting a fractional derivative based general formulation of the LScIS. These are described by fractional Euler–Cauchy equations. The fractional quantum derivatives are suitable for dealing with these systems. The use of the Mellin transform allowed us to define the multiplicative convolution and, from it, it is shown that the power function is the eigenfunction of the LScIS and the eigenvalue is the transfer function.

The computation of the impulse response from the transfer function is done following a procedure very similar to the used in the shift-invariant systems in Chap. 4. We will follow a two step procedure. In the first we solve a particular case with integer differentiation orders. Later we solve for the fractional case.

6.4.2 The General Formulation

We are going to consider a general formulation for the LScIS. The integer order case was studied in Yazici and Kashyap [9, 10]. To do it, we need the two fractional quantum derivatives that we presented in Sect. 6.2: the "below t" (analogue to anti-causal) and "above t" (analogue to causal) derivatives. If t were a time, we would talk on anti-causal and causal. We saw that working in the context of the Mellin transform we obtain two different regions of convergence: left and right relatively to a vertical straight line. This is not needed when dealing with integer order systems because we only have one Mellin transform for $t^n f^{(n)}(t)$ if n is integer. We rewrite here the two fractional quantum derivatives we are going to use

$$D_q^{\alpha} f(t) = \lim_{q\to 1} \frac{\sum_{j=0}^{\infty} \left[\begin{matrix}\alpha\\ j\end{matrix}\right]_q (-1)^j q^{j(j+1)/2} q^{-j\alpha} f(q^j t)}{(1-q)^{\alpha} t^{\alpha}} \tag{6.55}$$

and

$$D_{q^{-1}}^{\alpha} f(t) = \lim_{q\to 1} \frac{\sum_{j=0}^{\infty} \left[\begin{matrix}\alpha\\ j\end{matrix}\right]_q (-1)^j q^{j(j-1)/2} f(q^{-j} t)}{(1-q^{-1})^{\alpha} t^{\alpha}} \tag{6.56}$$

where $0 < q < 1$. When α is a positive integer, these derivatives lead to the results obtained by Yazici and Kashyap [9, 10]. We must give a special emphasis on an interesting fact: these derivatives are not local (unless α is positive integer), because they use infinite values on the left or on the right. So, the whole left or right history of the signal is needed. This is important in systems based on these derivatives: they exhibit long-range memory. With the adopted Mellin transform (6.4) we are led to results similar to those obtained with the Laplace transform in the study of shift invariant systems. The Mellin transforms of the above derivatives are given by (6.29) and (6.42):

$$\mathbf{M}\left[D_q^{\alpha} f(t)\right] = \frac{\Gamma(1+s+\alpha)}{\Gamma(1+s)} F(s+\alpha)$$

valid for $\mathrm{Re}(s) > -\min(0,\alpha) - 1$, in the first case and by

$$\mathbf{M}\left[D_{q^{-1}}^{\alpha} f(t)\right] = (-1)^{\alpha} \cdot \frac{\Gamma_q(-s)}{\Gamma_q(-s-\alpha)} F(s+\alpha)$$

valid for $\mathrm{Re}(s) < -\max(0,\alpha)$ in the second case. It is worth remarking that the first corresponds to the causal case when working in the Laplace transform context, while the second corresponds to the anti-causal one.

6.4.3 The Eigenfunctions and Frequency Response

We assume that the fractional LScIS is described by the general Euler–Cauchy differential equation

$$\sum_{i=0}^{N} a_i t^{\alpha_i} \cdot y^{(\alpha_i)}(t) = \sum_{i=0}^{M} b_i \cdot t^{\beta_i} \cdot x^{(\beta_i)}(t) \tag{6.57}$$

with $t \in R^+$. The response of the system is obtained by using the multiplicative convolution defined by (6.5). As said before the neutral element of this convolution is $g(t) = \delta(t-1)$. We must call the attention to the fact the point of application of the impulse is $t = 1$ and not $t = 0$, as it is the case of the shift-invariant systems. $\delta(t-1)$ is the inverse of $\Delta(s) = 1$. On the other hand, using the derivative definitions presented above, it is easy to show that:

$$[t^\alpha y^{(\alpha)}(t)] \vee g(t) = t^\alpha [y^{(\alpha)} f(t) \vee g(t)]$$

Let $h(t)$ be the impulse response of the system,

$$\sum_{i=0}^{N} a_i t^{\alpha_i} \cdot h^{(\alpha_i)}(t) = \sum_{i=0}^{M} b_i \cdot t^{\beta_i} \cdot \delta^{(\beta_i)}(t-1) \tag{6.58}$$

and convolve both sides of (6.58) with $x(t)$. We conclude immediately that

$$y(t) = \int_0^\infty h(t/u) x(u) \frac{du}{u} \tag{6.59}$$

If $x(t) = t^\sigma$, then

$$y(t) = H(\sigma) \cdot t^\sigma \tag{6.60}$$

meaning that the power function is the eigenfunction of the system described by (6.58) or (6.59) and $H(\sigma)$ is the eigenvalue, that we will call Transfer Function as in the shift-invariant systems and that is given by

$$H(s) = \int_0^\infty h(u) u^{-s-1} du \tag{6.61}$$

that is the Mellin transform of the impulse response. In (6.59) put $x(t) = g(at)$. It is a simple task to show that the output is $y(at)$ showing that the system is really scale invariant.

6.5 Impulse Response Computations

6.5.1 The Uniform Orders Case

Equation (6.57) is difficult to solve for any derivative orders. However, when the derivative orders have the format

$$\alpha_i = \alpha + i \quad i = 0, 1, 2, \ldots, N$$

and

$$\beta_i = \beta + i \quad i = 0, 1, 2, \ldots, N$$

we obtain a simpler equation

$$\sum_{i=0}^{N} a_i \, t^{\alpha+1} y^{(\alpha+i)}(t) = \sum_{i=0}^{M} b_i \, t^{\beta+1} x^{(\beta+i)}(t) \tag{6.62}$$

that we will solve with the help of the Mellin transform and using the fractional quantum derivatives. As we will show, the above equation allows us to obtain two transfer functions. Each of them has two terms that lead to two inverse functions. Before going into the general solution, we will consider the special integer order case with $\alpha = \beta = 0$

6.5.2 The Integer Order System with $\alpha = \beta = 0$

Consider a linear system represented by the differential equation

$$\sum_{i=0}^{N} a_i \, t^i \cdot y^{(i)}(t) = \sum_{i=0}^{M} b_i \cdot t^i x^{(i)}(t) \tag{6.63}$$

where $x(t)$ is the input, $y(t)$ the output, and N and M are positive integers $(M \leq N)$. Usually a_N is chosen to be 1. We will assume that this equation is valid for every $t \in R^+$. The system defined by (6.62) with $M = 0$ was already studied [see 9, 10]. However, it is interesting to repeat the computations here to acquire some background into the general case.

Applying the Mellin transform to both sides of (6.63) we obtain

$$\sum_{i=0}^{N} a_i (-1)^i (-s)_i \, Y(s) = \sum_{i=0}^{M} b_i \cdot (-1)^i (-s)_i \, X(s), \tag{6.64}$$

from where a transfer function is deduced

$$H(s) = \frac{Y(s)}{X(s)} = \frac{\sum_{i=0}^{M} b_i \cdot (-1)^i (-s)_i}{\sum_{i=0}^{N} a_i \cdot (-1)^i (-s)_i} \tag{6.65}$$

In this expression we need to transform both numerator and denominator into polynomials in the variable s. To do it we use the well known relation [11]

$$(x)_k = \sum_{i=0}^{k} (-1)^{k-i} v(k,i) x^i \tag{6.66}$$

where $v(,)$ represent the Stirling numbers of first kind that verify the recursion

$$v(n+1,m) = v(n,m-1) - nv(n,m) \tag{6.67}$$

for $1 \leq m \leq n$ and with
$v(n,0) = \delta_n$ and $v(n,1) = (-1)^{n-1}(n-1)!$
With some manipulation, we obtain:

$$\sum_{i=0}^{N} a_i (-1)^i (-x)_i = \sum_{i=0}^{N} \sum_{k=i}^{N} a_k (-1)^k v(k,i)\,(-x)^i = \sum_{i=0}^{N} A_i x^i \tag{6.68}$$

with the A_i coefficients given by

$$A_i = (-1)^i \sum_{i=k}^{N} a_k (-1)^k v(k,i) \tag{6.69}$$

or in a matricial format

$$\mathbf{A} = \mathbf{V} \cdot \mathrm{a} \tag{6.70}$$

where

$$\mathbf{A} = [A_0\, A_1 \ldots\ldots A_N]^T \tag{6.71}$$

$$\mathbf{V} = [v(i,j),\ i,j = 0, 1, \ldots, N] \tag{6.72}$$

and

$$\mathbf{a} = [a_0\, a_1 \ldots\ldots a_N]^T \tag{6.73}$$

With this formulation, the transfer function is given by:

$$H(s) = \frac{\sum_{i=0}^{M} B_i\, s^i}{\sum_{i=0}^{N} A_i\, s^i}\ M \leq N \tag{6.74}$$

that is the quotient of two polynomials in s. In the integer order case, it is indifferent which derivative we use, because they lead to the same result (6.17). This is a consequence of two facts:

(a) The derivatives (6.55) and (6.56) coincide with the classic when $\alpha = N \in Z^+$;
(b) The transforms defined in (6.29) and (6.42) are equal and the region of convergence is the whole complex plane

In general $H(s)$ has the following partial fraction decomposition

$$H(s) = \frac{B_M}{A_N} + \sum_{i=1}^{N} \sum_{j=1}^{m_i} \frac{a_{ij}}{(s - p_i)^j} \tag{6.75}$$

The constant term only exists when $M = N$ and its inversion gives a delta at $t = 1$:

$$\mathbf{M}^{-1}\left[\frac{B_M}{A_N}\right] = \frac{B_M}{A_N}\delta(t-1) \tag{6.76}$$

For inversion of a given partial fraction, we must fix the region of convergence $\mathrm{Re}(s) > \mathrm{Re}(p_i)$ or $\mathrm{Re}(s) > \mathrm{Re}(p_i)$ similar to identical situation found in the usual shift invariant systems with the Laplace transform. Let us assume that the poles are simple. From the inversion Mellin integral, we obtain [3]

$$\mathbf{M}^{-1}\left[\frac{1}{(s-p)}\right] = w(t) \cdot t^p \tag{6.77}$$

where $w(t)$ is equal to $u(1-t)$ or to $u(t-1)$ in agreement with the adopted the region of convergence. By successive derivation in order to p we obtain the solution for higher order poles

$$\mathbf{M}^{1}\left[\frac{1}{(s-p)^k}\right] = w(t)\,\frac{(-1)^{k-1}[\log(t)]^{k-1}}{(k-1)!}t^p \tag{6.78}$$

We conclude that the response corresponding to an input $\delta(t-1)$ is given by:

$$h(t) = \frac{B_M}{A_N}\delta(t-1) + \sum_{i=1}^{N}\sum_{k=1}^{m_i} a_{ik} \cdot \frac{(-1)^{k-1}[\log(t)]^{k-1}}{(k-1)!}t^{p_i}w(t) \tag{6.79}$$

To compute the output to any function $x(t)$ we only have to use the multiplicative convolution. As in the shift-invariant systems, we have several ways of choosing the region of convergence. We can have all right signals, all left signals or, mixed right and left signals. In [9, 10] the first term does not appear, since only the all-pole case was discussed.

It is interesting to make here an important remark. Verify that (6.79) behaviours like the usual responses of the anti-causal and causal systems. When $\mathrm{Re}(p_i) > 0$ and $t > 1$, it increases without bound as $t \to \infty$, while it decreases as $t \to 0$. if $\mathrm{Re}(p_i) < 0$, (6.79) increases without bound as $t \to 0$, while it decreases as $t \to \infty$. This means that we can use the well known Routh–Hurwitz test to study the stability of LScIS.

6.5.3 The Fractional Order System

Consider now a linear system represented by the fractional differential equation

$$\sum_{i=0}^{N} a_i\, t^{\alpha+i} \cdot y^{(\alpha+i)}(t) = \sum_{i=0}^{M} b_i \cdot\ t^{\beta+i} x^{(\beta+i)}(t) \tag{6.80}$$

where α and β are positive real numbers. With the Mellin transform we obtain two different transfer functions depending on the derivative we use, (6.55) or (6.56). Using derivative (6.55) and its Mellin transform we have:

$$H(s) = \frac{\sum_{i=0}^{M} b_i(-1)^i(s-\beta)_i}{\sum_{i=0}^{N} a_i(-1)^i(s-\alpha)_i} \cdot \frac{\Gamma(1+s-\alpha)}{\Gamma(1+s)} \frac{\Gamma(1+s)}{\Gamma(1+s-\beta)} \tag{6.81}$$

Proceeding as in (6.5.2) we have

$$H(s) = \frac{\sum\limits_{i=0}^{M} B_i(s-\beta)^i}{\sum\limits_{i=0}^{N} A_i(s-\alpha)^i} \cdot \frac{\Gamma(1+s-\alpha)}{\Gamma(1+s-\beta)} \tag{6.82}$$

So, the transfer function in (6.82) has two parts, the first is similar to (6.74) aside translations on the pole and zero positions. Its inverse has the format:

$$h(t) = \frac{B_M}{A_N}\delta(t-1) + t^{\alpha} \sum_{i=1}^{N} \sum_{k=1}^{m_i} c_{ik} \cdot \frac{(-1)^{k-1}[\log(t)]^{k-1}}{(k-1)!} t^{p_i} w(t) \tag{6.83}$$

where $\alpha + p_i, i = 1, 2, \ldots, N$ are the poles. We must remark that it does not depend explicitly on β. The second factor in (6.82) leads to a new convolutional factor needed to compute its complete inversion. So, we have to compute the inverse Mellin transform of

$$H_a(s) = \frac{\Gamma(1+s-\alpha)}{\Gamma(1+s-\beta)} \tag{6.84}$$

For taking account with the stability of the system, we can consider the region of convergence the half plane defined by $\mathrm{Re}(s) > 0$. This function has infinite poles at $s = \alpha - 1 - n$, with n a non negative integer. To invert it we can always choose an integration path on the right of all the poles similarly to the path shown in Fig. 6.1, but with the most left segment infinitely far. The residues are given by

$$R_n = \frac{(-1)^n t^{\alpha-1-n}}{\Gamma(\alpha-\beta-n)!} u(t-1)$$

according to the properties of the Gamma function [7]. Adding the residues, we obtain

$$h_a(t) = \frac{t^{\alpha-1}}{\Gamma(a-\beta)} \sum_{n=0}^{\infty} \frac{\Gamma(a-\beta)(-1)^n t^{-n}}{\Gamma(a-\beta-n)n!} u(t-1)$$

where we can identify the binomial series. Summing it, we obtain

$$h_a(t) = \frac{1}{\Gamma(\alpha-\beta)} t^{\beta}(t-1)^{\alpha-\beta-1} u(t-1) \tag{6.85}$$

So, the impulse response corresponding to (6.82) is the multiplicative convolution of (6.83) and (6.85). However, we can obtain an alternative approach to invert (6.82). It consists in expanding its first term in N partial fractions and invert N transforms with the format $\frac{\Gamma(1+s-\alpha)}{(s-\alpha-p)\Gamma(1+s-\beta)}$. By simplicity, we assumed that all the poles are simple. We proceed as above to compute the residues. Collecting them the impulse response is given by

$$\begin{aligned} h(t) = & \frac{B_M 1}{A_N \Gamma(\alpha-\beta)} t^{\beta}(t-1)^{\alpha-\beta-1} u(t-1) + t^{\alpha} \sum_{i=1}^{N} C_i \cdot \frac{\Gamma(1+p_i)}{\Gamma(\alpha-\beta+p_i+1)} t^{p_i} u(t-1) \\ & - \frac{(-1)^{\beta-\alpha+1}}{\Gamma(\alpha-\beta)} t^{\beta} \sum_0^{\infty} \binom{\alpha-\beta-1}{n} (-1)^n \frac{t^n}{\beta-\alpha-p_i+n} u(t-1) \end{aligned} \tag{6.86}$$

Choosing the other derivative (6.56) and its Mellin transform (6.45), we have

$$H(s) = \frac{\sum_{i=0}^{M} B_i(s-\beta)^i}{\sum_{i=0}^{N} A_i(s-\alpha)^i} \cdot (-1)^{\beta-\alpha} \frac{\Gamma(-s+\beta)}{\Gamma(-s+\alpha)} \tag{6.87}$$

The first factor has as inverse the expression given by (6.83) for $w(t) = -u(1-t)$. For the second we proceed as before. Now the integration path is in the right half complex plane as in Fig. 6.2 but with the most right segment infinitely far.

We proceed as above to obtain

$$h_a(t) = \frac{1}{\Gamma(\alpha-\beta)} t^{\beta}(t-1)^{\alpha-\beta-1} u(1-t) \tag{6.88}$$

To compute the final impulse response we only have to proceed as in the other case. We obtain, for the simple pole case

$$\begin{aligned} h(t) = & \frac{B_M 1}{A_N \Gamma(\alpha-\beta)} t^{\beta}(t-1)^{\alpha-\beta-1} u(1-t) + (-1)^{\beta-\alpha+1} t^{\alpha} \sum_{i=1}^{N} C_i \cdot \frac{\Gamma(\beta-\alpha+p_i)}{\Gamma(p_i)} t^{p_i} \cdot u(1-t) \\ & + \frac{(-1)^{(\beta-\alpha+1)}}{\Gamma(\alpha-\beta)} t^{\beta} \sum_0^{\infty} \binom{\alpha-\beta-1}{n} (-1)^n \frac{t^n}{\beta-\alpha-p_i+n} u(t-1) \end{aligned} \tag{6.89}$$

We must remark that the above results are valid even if α and β are positive integers. Of course, we could obtain other solutions by choosing other integration paths such that there were poles on the left and on the right of it. In these cases we would obtain "two-sided" responses. It is interesting to remark that:

Fig. 6.1 Integration path for the inverse Mellin transform of $\frac{\Gamma(1+s-\alpha)}{\Gamma(1+s-\beta)}$ with $\mathrm{Re}(s) > 0$

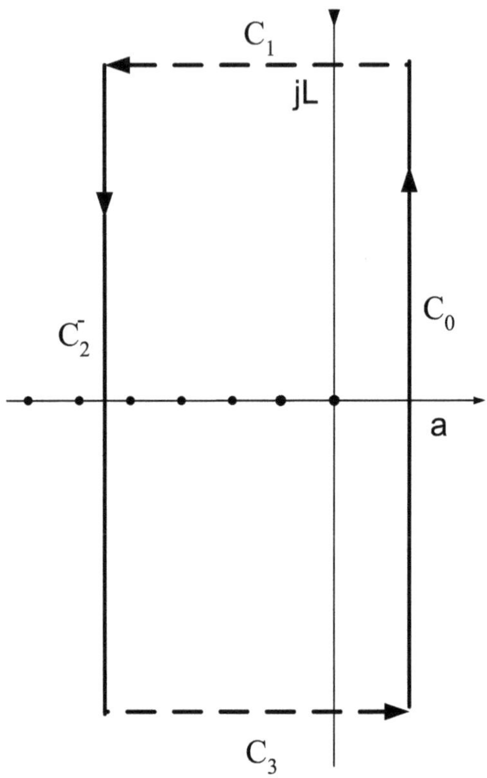

If $\alpha = \beta$, the second terms in (6.82) and (6.87) are equal to 1, implying that the complete impulse response is given by (6.83).

When $\alpha = 0$ and $\beta \neq 0$ in (6.86) we obtain a situation very similar to the one treated by Yazici and Kashyap [9, 10].

If $\alpha = \beta + 1$, (6.85) and (6.88) become merely power functions and so self-similar.

6.5.4 A Simple Example

We are going to consider a simple system described by the differential equation:

$$t^{\alpha+1} y^{(\alpha+1)}(t) + a\, t^{\alpha} y^{(\alpha)}(t) = x(t)$$

If $\alpha = 0$, the impulse response comes from (6.83) and it is given by:

$$h_s(t) = t^{-a} w(t)$$

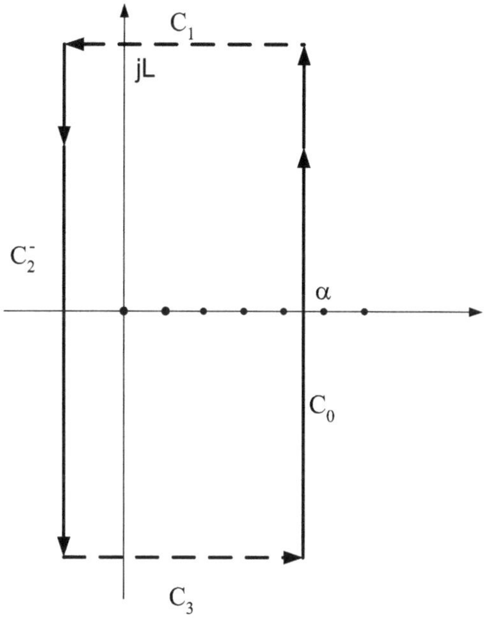

Fig. 6.2 Integration path for the inverse Mellin transform of $(-1)^{\beta-\alpha}\frac{\Gamma(-s+\beta)}{\Gamma(-s+\alpha)}$ with $h\,\mathrm{Re}(s) < 0$

where $w(t)$ is equal to $-u(1-t)$ or to $u(t-1)$, in agreement with the adopted region of convergence. The analogue shift invariant corresponding system

$$y'(t) + a\,y(t) = x(t)$$

has the causal and anti-causal impulse responses:

$$h_t(t) = \pm e^{-at}u(\pm t)$$

As seen, we made a substitution $t \to e^t$.

Now, let $\alpha \neq 0$. We have, from (6.86)

$$\begin{aligned} h(t) = {} & \frac{\Gamma(1-a)}{\Gamma(\alpha - a + 1)} t^{\alpha-a} u(t-1) \\ & + \frac{1}{\Gamma(\alpha)} t^{\alpha-1} \sum_{0}^{\infty} \binom{\alpha-1}{n} (-1)^n \frac{t^{-n}}{-a+n+1} u(t-1) \end{aligned} \tag{6.90}$$

and, from (6.89)

$$\begin{aligned} h(t) = {} & (-1)^{-\alpha+1} \frac{\Gamma(-\alpha-a)}{\Gamma(-a)} t^{\alpha-a} u(1-t) \\ & + \frac{(-1)^{-\alpha+1}}{\Gamma(\alpha)} \sum_{0}^{\infty} \binom{\alpha-1}{n} (-1)^n \frac{t^n}{a-\alpha+n} u(1-t) \end{aligned}$$

The shift invariant corresponding system

$$y^{(\alpha+1)}(t) + a\,y^{(\alpha)}(t) = x(t)$$

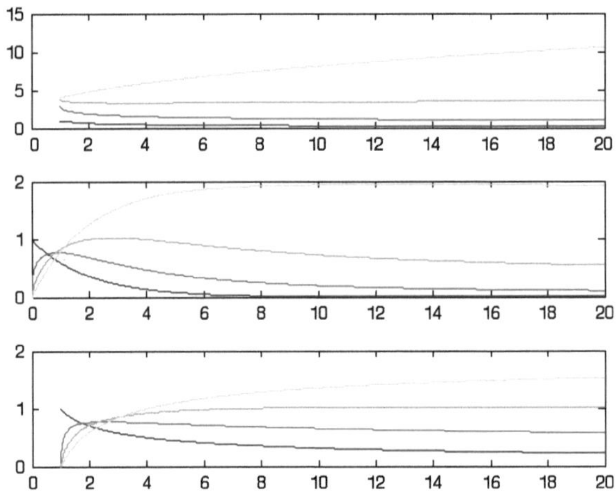

Fig. 6.3 Impulse responses obtained with (6.90) and (6.91) for $\alpha = 0$, 0.33, 0.66, and 0.99 (strips uppermost and middle). In the last it is shown the result of a $\log(t)$ transformation applied to the functions in the middle strip

has the following transfer function

$$H(s) = \frac{s^{-\alpha}}{s+a}$$

and its causal impulse response is (see Chap. 4):

$$h_s(t) = \sum_0^{\infty} (-a)^n \frac{t^{n+\alpha}}{\Gamma(n+\alpha+1)} u(t) \tag{6.91}$$

As seen the above referred substitution seems not to be valid here. The anti-causal response is very similar, but it does not have any special interest. In Fig. 6.3, we present the results obtained for these systems, for several values of $\alpha\,(0, 0.33, 0.66, 0.99)$. The upper strip shows the results obtained with (6.90). The results in the middle strip were obtained with (6.91). The third strip shows the result of a transformation $t \to \log(t)$ in (6.91). Although it is not very clear in the picture, we can see the similarity between the curves corresponding to $\alpha = 0$ and $\alpha = 0.99$ with the equivalent in the upper strip.

6.5.5 Additional Comments

The impulse responses stated (6.86) and (6.89) depend directly on the differential equation (6.62) not on the way we followed to obtain them. This means that we are not obliged to use the quantum derivative. In fact we could also use another derivative like Grunwald–Letnikov, Riemann–Liouville or Caputo, but it would be very difficult to arrive at the results we obtained. The quantum derivative allows us to obtain such impulse responses more easily. On the other hand, those derivatives are suitable for dealing with shift-invariant systems defined over R, not R^+.

In the integer order case, we can switch from the LScIS to the corresponding linear shift-invariant systems: we only have to perform a logarithmic transformation. However, this is not evident neither correct in the fractional case, due to the first term (6.86) and (6.89), as the example presented above shows. This fact may come from the difficulty in defining fractional derivative of a composite function. The lack of emphasis on this fact is due to the desire of presenting a linear system that exists by itself and not because can be the transformation of another one. It is more or less the same situation that we find when introducing difference equations. They exist and do not need to be presented as transformations of the ordinary differential equations (with bilinear or other mapping). It is curious to refer that we can obtain the corresponding shift invariant system, by considering that the transfer function in (6.62) is now a transfer function of a shift invariant system and use the Laplace transform to go back into a new differential equation.

The LScIS, being scale invariant, but not shift invariant, can be useful in detection problems and in image processing. Their conjunction with the Wavelet transform can be interesting [9, 10].

6.6 Conclusions

We presented the quantum fractional derivative as an alternative to the common Grünwald–Letnikov and Liouville derivatives. It was described in two formulations: summation and integral. Its Mellin transform was also presented and use to establish the relation between the two formulations. The summation formulations are similar to the Grünwald–Letnikov fractional derivatives. The main difference lies in the use of an exponential scale for the independent variable. The Grünwald–Letnikov derivatives use a linear scale. The integral formulations are similar to the Liouville derivatives. This derivative is useful to solve fractional Euler–Cauchy differential equations and can be useful in dealing with scale problems.

We introduced the general formulation of the linear scale invariant systems through the fractional Euler–Cauchy equation. To solve this equation we used the fractional quantum derivative concept and the help of the Mellin transform. As in the linear time invariant systems we obtained two solutions corresponding to the use of two different regions of convergence. We presented other interesting features of the LScIS, namely the frequency response. We made also a brief study of the stability.

There is another way of introducing two-sided quantum derivatives. To do it, we can start from the two-sided quantum derivative

$$D_{q_0} f(t) = \lim_{q \to 1} \frac{f\left(q^{-1/2} t\right) - f\left(q^{1/2} t\right)}{\left(q^{-1/2} - q^{1/2}\right) t} \tag{6.92}$$

and proceed as in Chap. 5. It will not be done here.

References

1. Kac V, Cheung P (2002) Quantum calculus. Springer, New York
2. Ash JM, Catoiu S, Rios-Collantes-de-Terán R (2002) On the nth quantum derivative. J Lond Math Soc 2(66):114–130
3. Bertrand J, Bertrand P, Ovarlez JP (2000) The mellin transform. In: Poularikas AD (ed) The Transforms and Applications Handbook, 2nd edn. CRC Press, Boca Raton
4. Poularikas AD (ed) (2000) The transforms and applications handbook. CRC Press, Boca Raton
5. Koornwinder TH (1999) Some simple applications and variants of the q-binomial formula. Informal note, Universiteit van Amsterdam
6. Al-Salam W (1966) Some fractional q-integrals and q-derivatives. Proc Edin Math Soc 15:135–140
7. Henrici P (1991) Applied and computational complex analysis, vol 2. Wiley, New York, pp 389–391
8. Braccini C, Gambardella G (1986) Form-invariant linear filtering: theory and applications. IEEE Trans Acoust Speech Signal Process ASSP-34(6):1612–1628
9. Yazici B, Kashyap RL (1997) Affine stationary processes with applications to fractional Brownian motion. In: Proceedings of 1997 International Conference on Acoustics, Speech, and Signal Processing, vol 5. Munich, Germany, pp 3669–3672, April 1997
10. Yazici B, Kashyap RL (1997) A class of second-order stationary self-similar processes for 1/f phenomena. IEEE Trans Signal Process 45(2):396–410
11. Abramowitz M, Stegun I (1972) Stirling Numbers of the First Kind. Sect. 24.1.3 in Handbook of Mathematical Functions with Formulas, Graphs, and Mathematical Tables, 9th printing. Dover, New York, p 824

Chapter 7
Where are We Going To?

7.1 Some Considerations

The non-integer order systems can describe dynamical behavior of materials and processes over vast time and frequency scales with very concise and computable models. Nowadays well known concepts are being extended to the development of fractional robust control systems, signal filtering, identification and modelling. Of particular interest is the fact that the fractional systems exhibit both short and long term memory (in some areas the designation "long range processes" is firmly established). While the short term memory corresponds to the "distribution of time constants" associated with the distribution of isolated poles and zeroes in the complex plane, the long term memory corresponds to infinitely many interlaced close to each other poles and zeros that in the limit originate a branch cut line as we saw in Chap. 3. This translates to a lack of specific time scale and, therefore, no new resonance or other instability effects appear and incorporates the power law behavior found in natural systems that show the greatest robustness to variation of environmental parameters. These characteristics have great influence on the development and application of fractional systems that is dependent on satisfactory solutions for the traditional tasks: modelling, identification, and implementation. In the fractional case, they are slightly more involved due to the fact of having, at least, one extra degree of freedom: the fractional order. However, this difficult increments the possibilities of obtaining more reliable and robust systems. This is challenging and people working in the area has been giving different interesting answers. We can refer the following approaches:

- Circuit implementations with fractional elements—it consists of using the classic circuit theory, but with fractional capacitors and coils.
- Trans-finite circuits—the infinite transmission lines are circuits with fractional behaviour, but there are other interesting circuits with similar characteristics like

M. D. Ortigueira, *Fractional Calculus for Scientists and Engineers*,
Lecture Notes in Electrical Engineering, 84, DOI: 10.1007/978-94-007-0747-4_7,

the tree fractance (a tree of RC circuits) and chain fractance (a series of parallel RC) circuits.
- Band-limited approximations—it is an engineer approach. There are several ways of doing the design and implementation we can refer (a) the CRONE that uses the Bode diagrams band (b) the continued fraction approaches. Both construct pole-zero systems with interlaced poles and zeros.
- Identification from frequency data—it consists on a least-squares approach in the frequency domain. The more interesting algorithm uses a generalized Levy method.
- Discrete-time implementations—there are several algorithms that start from an s to z conversion and design an ARMA model.
- Weighted summation of exponentials—it consists in approximating the impulse response of the system through a weighted summation of exponentials by which as the number of elements increases toward infinity describes fractional behaviour.

7.2 Some Application Examples

7.2.1 The Age of the Earth Problem

Normally the solution of the tautochrone problem by Abel is considered the first application of the fractional derivative, although Abel was unaware of this fact. This is described in a lot of books, papers and Internet sites. So, we will not describe it. Alternatively we will offer a brief description of a curious application by Heaviside [1].

In the last 35 years of the XIXth century the Earth Age controversy involved several great scientists, including Lord Kelvin that put the upper bound in 98 million years. Heaviside assumed that the Earth could be considered as a semi-infinite mass filling all the space for $x \geq 0$ at an initial zero temperature. Suddenly a step elevated the temperature at surface to a given value Θ_0. The temperature gradient is infinite at $t = 0$, but will decrease to zero as the Earth will heat. He looked for the time interval needed to reach the observed gradient (aside a sign change). He started from the one-dimensional heat equation

$$\frac{\mathrm{d}\theta}{\mathrm{d}t} = \frac{k}{c}\frac{\partial^2 \theta}{\partial x^2} \tag{7.1}$$

$\theta(x, t)$ is the temperature at the distance x inside the Earth, c is the heat capacity and k is the thermal conductivity of the Earth. Applying the LT to (7.1) we obtain:

$$\frac{\partial^2 \Theta}{\partial x^2} = s\Theta\frac{c}{k} \tag{7.2}$$

where $\Theta(x,s)$ is the LT of $\theta(x,t)$ relatively to t. As the Earth is semi-infinite, suitable evident boundary conditions lead to the solution of the above equation:

$$\Theta(x,s) = \Theta_0 \cdot e^{-\sqrt{cs/kx}} \tag{7.3}$$

We must refer that Heaviside did not use the LT, but his operator p that he treated as a number. The LT of temperature gradient, $G(0, s)$, at the surface is

$$G(0,s) = \left(\frac{\partial\Theta}{\partial x}\right)_{x=0} = \Theta_0\sqrt{\frac{cs}{k}} \tag{7.4}$$

He considered that this was an operator acting on the input, a step, with LT $1/s$. Computing the inverse LT, we obtain

$$g(0,t) = \Theta_0\sqrt{\frac{c}{k}}\frac{t^{-1/2}}{k\Gamma(1/2)}u(t) = T_0\sqrt{\frac{c}{\pi k}}t^{-1/2}u(t) \tag{7.5}$$

Heaviside admitted that if the earth could cool from the state back toward zero it would take the same time to reach the observed surface gradient (aside a sign). Then the age of the Earth should be given by:

$$T = \Theta_0^2\frac{c}{\pi k g^2(0,T)} \tag{7.6}$$

With $\Theta_0 = 3,900°C$, $g(0, T) = 1°C$ per 2,743 cm, and $k/c = 0.01178\ cm^2/s$, he obtained Kelvin's result. This approach was modified by Perry in order to become more realistic, but maintaining the essential of Heaviside mathematical formulation. With it, Perry obtained the value $T = 315$ million years.

7.3 Biomedical Applications

7.3.1 General Considerations

The first applications of fractional calculus to biomedical problems were done in the areas of membrane biophysics and polymer viscoelasticity, where the experimentally observed power law dynamics for current–voltage and stress–strain relationships were concisely captured by fractional order differential equations. On the other hand, there is evidence that biological signals (ECG, EMG, and EEG) have spectra that do not increase or decrease by multiples of 20 dB. Hence, fractional system models are often proposed for both analytical and empirical reasons. Here, we consider examples of biomedical applications of fractional calculus taken from the fields of bioinstrumentation, mechanobiology and biomedical imaging.

Physiological models based on linear differential equations are highly successful in describing a wide range of complex phenomena (e.g., action potential propagation, blood oxygenation and filtration, and feedback control of insulin secretion). Such models, also serve as the basis for understanding normal physiological homeostatis, as well as the changes that arise as a consequence of disease. Physiological models connect events at the molecular level (ion transport, gas diffusion, vesicle formation) to those at the organ level (blood clearance, oxygen uptake/gram tissue, muscle tension). Much current work in biophysics and physiology is directed toward linking molecular processes with whole organ (brain, heart, and muscle) function by developing multiscale models that span the intermediate levels of structure (e.g., from the centimeter dimensions of gross anatomy down to the submicron resolution of histology). In building multiscale models one can either try to use as much of the available anatomical and histological knowledge as possible—building highly complex structures with hundreds of components (organelles, membranes, cells, extracellular matrix, etc.)—or try to deal empirically with the complexity by developing whole system descriptions (e.g., linear, non-linear, deterministic, or stochastic models) with embedded chaotic or fractal measures (fractal dimensions, Lyapunov exponents, non-Gaussian probability distributions) that capture important features of the observed behavior.

7.3.2 Fractional Dynamics Model

A fractional order model is commonly used to describe the behavior of neural systems (sensory and motor). A simple example is the vestibular–oculomotor system modeled by Anastasio in the Laplace domain as s^{α}, where $-1<\alpha<1$. The occurrence of s^{α} behavior in the transfer functions for the neural components of vestibulo–oculomotor systems suggests its need to control or monitor the underlying biological, physical, or chemical mechanisms. The s^{α} behavior follows directly from observed power law transient and dynamic behavior unique to the anatomical structure or neurological connections of living systems. Thus, the subthreshold behavior of axons, which mimic at their most basic level lossy (RC) transmission lines with fractional impedance relationships, could play a role in understanding synapse complexity, dendritic convergence and generator potential initiation.

For example, the encoding of head motion by the inner ear arises via convergence of unmyelinated afferent and efferent nerve fibers in the vestibular neuroepithelium. This has been suggested as an anatomical site where summation of excitatory and inhibitory postsynaptic potentials can occur. In a paper on distributed relaxation processes in sensory adaptation, John Thorson and Marguerite Biederman-Thorson reviewed earlier interpretations for fractional dynamics (nonlinear spring, transmission line, and Gaussian distribution of exponential rate constants), which they found for the most part, to provide an incomplete explanation for the wide dynamic range of sensory adaptations. These considerations led to a fractional order model.

7.3.3 Fractional Impedance Model

Distributed relaxation processes appear to be common in cells and tissues. Therefore, it should not be surprising to see that fractional calculus can play an important role in describing the input–output behavior of biological systems. The physical foundations for this behavior may be sought in the fractal or porous structure of the system components or in the physical characteristics of its surfaces and interfaces. Much work is ongoing to develop a direct link between fractal models of molecules, surfaces, and materials and the fractional kinetics or dynamics of the resulting behavior (polymerization electrochemical reactions, viscoelastic relaxation).

Fractional order circuit elements, such as the impedance: $Z = Z_0/(s)^\alpha$ or $Z = Z_0/(j\ \omega)^\alpha$, where $0<\alpha<1$, provide a useful model for describing the frequency response of dielectrics and biological tissues. Such circuit elements can also be used to develop an electrical circuit model of the electrode–cardiac tissue interface of a pacemaker electrode [2]. Accurate impedance models are essential for designing cardiac pacemakers. Fractional calculus appears in the model through a fractional order (or constant phase, $Z = Z_0\omega^{-\alpha}\exp(j\tan^{-1}(\pi\alpha/2))$) circuit element that governs diffusion limited electrochemical reactions at the surface of the electrode.

We can use the correspondence between RC electric circuits and viscoelastic networks of springs and dashpots to construct similar fractional order dynamic models for the biomechanical properties of tissues. For example, Craiem and Armentano have modelled the elastic properties of the aorta, in vivo in a Merino sheep, using a fractional order generalization of the relationship between stress $\sigma(t)$and strain $\varepsilon(t)$. Their generalized Voigt model consists of a spring in parallel with two "springpots" of fractional order.

Fractional order models have also been used by [3] to fit magnetic resonance elastography (MRE) data from breast tumors. In this technique, MRI is used to image low frequency (50–1,500 Hz) shear wave oscillations in the breast. The wavelength and attenuation of the vibrations directly reflect the elastic shear modulus and the viscosity of the tissue through a complex wave vector: $k(\omega) = \beta(\omega) + j\alpha(\omega)$. In MRE these tissue properties are mapped into an elastogram image through an assumed model of the tissue's mechanical properties—usually a purely elastic spring with zero loss, or a Voigt spring/dashpot model. In his study, Sinkus assumed a power law increase in attenuation with excitation frequency, $\alpha(\omega) = \alpha_0\omega^y$ (where $0 < y < 1$).

In the three examples considered here, fractional order models were found to provide better fits to electrical and mechanical measurements made on living tissue. Such studies need replication, but these findings provide useful examples of cases where an extension of the "standard" integer order dynamic models of circuits and mechanical systems is warranted. Fractional order dynamic models of complex, multiscale systems account for anomalous dynamic behavior through a simple extension of the order of the operations from integer to fractional. In the time domain this extension is manifest through incorporation to a variable degree of system memory through convolution with a power law kernel exhibiting fading memory of the past. Perhaps, in the future, the development of integrated space and time domain fractional order models will become a more standard component of linear systems analysis, at least when such models are applied to living systems. Clearly, when the structure in living systems is fractal, or when the measured signals exhibit anomalous properties, one should suspect that such dynamics might best be expressed by fractional order models.

7.3.4 Additional Comments

Fractional calculus models provide a relatively simple way to describe the physical and electrical properties of complex, heterogeneous, and composite biomaterials. There is a multi-scale generalization inherent in the definition of the fractional derivative that accurately represents interactions occurring over a wide range of space or time. Thus, we can avoid excessive segmentation or compartmentalization of tissues into subsystems or subunits—a system reduction that often creates more computational and compositional complexity than can be experimentally evaluated. Finally, fractional calculus models suggest new experiments and measurements that can shed light on the meaning of biological system structure and dynamics. Thus, by applying fractional calculus to model the behavior of cells and tissues, we can begin to unravel the inherent complexity of individual molecules and membranes in a way that leads to an improved understanding of the overall biological function and behavior of living systems.

7.4 The Fractional Brownian Motion

Fractional Brownian motion was introduced first by [4]. Later, Mandelbrot and Van Ness proposed it as a model for non-stationary signals, with stationary increments, that are useful in understanding phenomena with long range dependence and with a frequency dependence of the form $1/f^{\alpha}$, with α non-integer. To introduce it in the context of the fractional derivative we can proceed as follows [5, 6].

Assume that we are computing the fractional derivative of the white noise, $w(t)$, with power equal to σ^2. We define a fractional noise by:

$$r_\alpha(t) = D^\alpha w(t) \tag{7.7}$$

If $w(t)$ is Gaussian, we will call $r_\alpha(t)$ fractional Gaussian noise. In the interval $-1/2<\alpha<1/2$ we obtain a stationary process, in the anti-derivative case ($\alpha<0$), and non-stationary, in the derivative case ($\alpha > 0$). This fractional noise will be used next to define the fractional Brownian motion. Let $r_\alpha(t)$ be a fractional noise. Define a process $v_\alpha(t)$, $t\geq 0$, by:

$$v_\alpha(t) = \int_0^t r_\alpha(\tau)\,\mathrm{d}\tau \tag{7.8}$$

We will call this process a *fractional Brownian motion* (or generalised Wiener–Lévy process). We can show that it enjoys all the properties normally required for the fBm. When $\alpha = 1$, we recover the usual Brownian motion.

7.5 Future Travels

In the previous sections we made brief descriptions of systems and situations where the fractional behaviour is clear and its use very beneficial. As said fractional derivatives are suitable for understanding an increasing number of physical, biological, economical and social phenomena. This may lead us to infer that it will invade other areas of human activity. Much remains to be done, and we look the philosopher Henri Bergson to provide inspiration, for, as he noted in his 1911 work *Creative Evolution*: "the present contains nothing more than the past, and what is found in the effect was already in the cause". Prof. Nishimoto produced the prophetical affirmation: "The Fractional Calculus will be the XXIth century calculus". In the same line of reasoning we can say that "the fractional systems will be the systems of the XXIth century systems" and the recent developments and applications reinforce this assertion. Even today it would be almost impossible to give a complete picture of the present state of the art concerning the application areas. Surely these will be enlarged to include others. The technological and scientific achievements will determine the speed of introduction of fractional calculus in other areas.

We believe that the fractional calculus is ready for use in all aspects of Signals and Systems. What is necessary for researchers is to have access to the important tools of the theory. This was one objective of this text: to introduce the fractional linear systems and offer some insights into how the somehow involved mathematics is applied to very practical problems.

And what about the fractional calculus itself? A lot of developments were introduced in the last 20 years, but we need much more, e.g.

(a) Continuous-time to discrete-time conversion preserving the long memory of the systems.
(b) Study of the interaction between the fractional systems and the stochastic processes.
(c) Establishment of a bridge between the fractional signals obtained from fractional systems and the alpha stable processes.
(d) Study of distributed systems involving the causal and acausal (two-sided) derivatives.
(e) Continuation of the study of the initial condition problem.

(f) Study of the quantum derivative its applications as well as the scale invariant systems.
(g) Physical and/or geometrical meaning of a fractional derivative.
(h) …

GOOD TRAVEL!

References

1. Nahim PJ (1985) Oliver Heaviside, fractional operators, and the age of the Earth. IEEE Trans Education 28(2):94–104
2. Ovadia M, Zavitz DH (1998) The electrode–tissue interface in living heart: equivalent circuit as a function of surface area. Electroanalysis 10:262–272
3. Sinkus R, Siegmann K, Xydeas T, Tanter M, Claussen C, Fink M (2007) MR elastography of breast lesions: understanding the solid/liquid duality can improve the specificity of contrast-enhanced MR mammography. Magn Reson Med 58:1135–1144
4. Kolmogorov NA (1940) Wienersche Spiralen und Einige Andere Interessante Kurven im Hilbertschen Raum. Compte Rendus de l'Académie des Sciences de l'Union des Républiques Soviétiques Socialistes 26(2):115–118
5. Ortigueira MD, Batista AG (2004) A fractional linear system view of the fractional brownian motion. Nonlinear Dynamics 38:295–303
6. Ortigueira MD, Batista AG (2008) On the relation between the fractional Brownian motion and the fractional derivatives. Phys Lett A 372(7):958–968

MIX
Papier aus verantwortungsvollen Quellen
Paper from responsible sources
FSC® C105338

If you have any concerns about our products, you can contact us on
ProductSafety@springernature.com

In case Publisher is established outside the EU, the EU authorized representative is:
Springer Nature Customer Service Center GmbH
Europaplatz 3, 69115 Heidelberg, Germany

Printed by Libri Plureos GmbH
in Hamburg, Germany